MW01701741

The Unofficial

Ages 7+

MINECRAFT MATH WORKBOOK

2-digit and 3-digit Addition & Subtraction (B)

STEM mindset, LLC 1603 Capitol Ave. Suite 310 A293 Cheyenne, WY 82001 USA
www.math-stepbystep.com info@math-stepbystep.com

First published in the USA 2019. ISBN 9781948737500

Contents

1	A	10	J	19	S
2	B	11	K	20	T
3	C	12	L	21	U
4	D	13	M	22	V
5	E	14	N	23	W
6	F	15	O	24	X
7	G	16	P	25	Y
8	H	17	Q	26	Z
9	I	18	R	A is the 1st, Z is the 26th	

CHAPTER 1

Fun with and without Renaming/Regrouping

Adding 2-Digit Numbers Vertically up to 1000

Adding 2-Digit Numbers Horizontally up to 1000

Subtracting 2-Digit Numbers Vertically

Subtracting 2-Digit Numbers Horizontally

Adding 3-Digit Numbers Horizontally up to 1000

Subtracting 3-Digit Numbers Horizontally

Number Talks

Olympiad Math Problems

CogAT test prep

Puzzles

Mazes

Word Search

Cryptograms

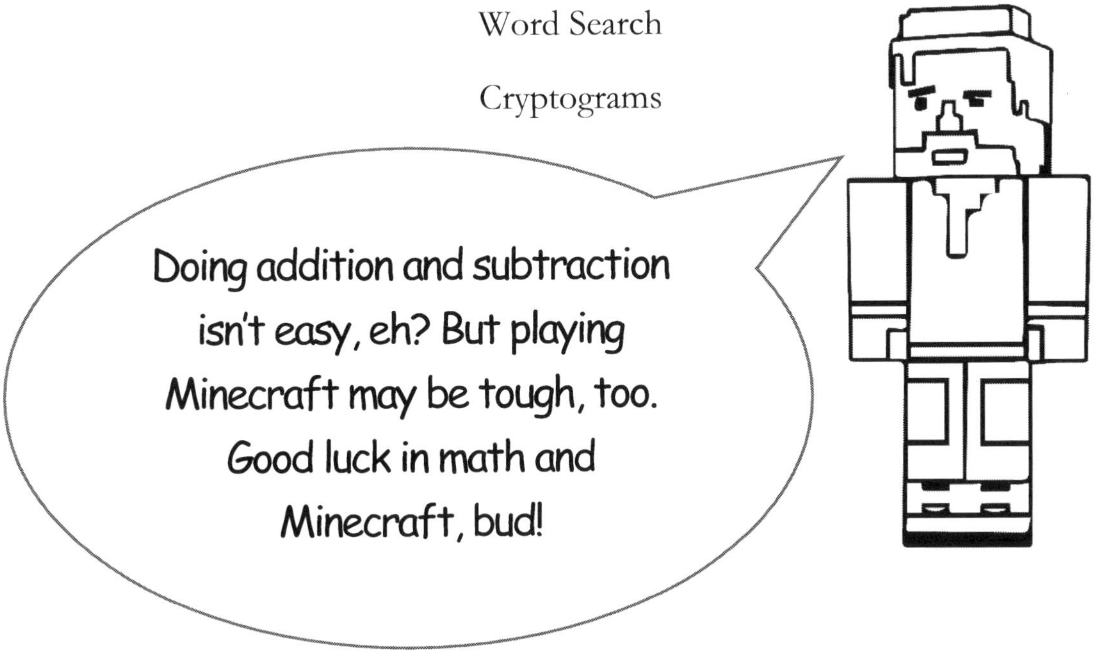

1. Add or subtract.

50 – 6 = __ 60 – 3 = __ 80 – 7 = __

72 + 8 = __ 87 + 3 = __ 46 + 4 = __

2. Subtract. Decompose the subtrahend.

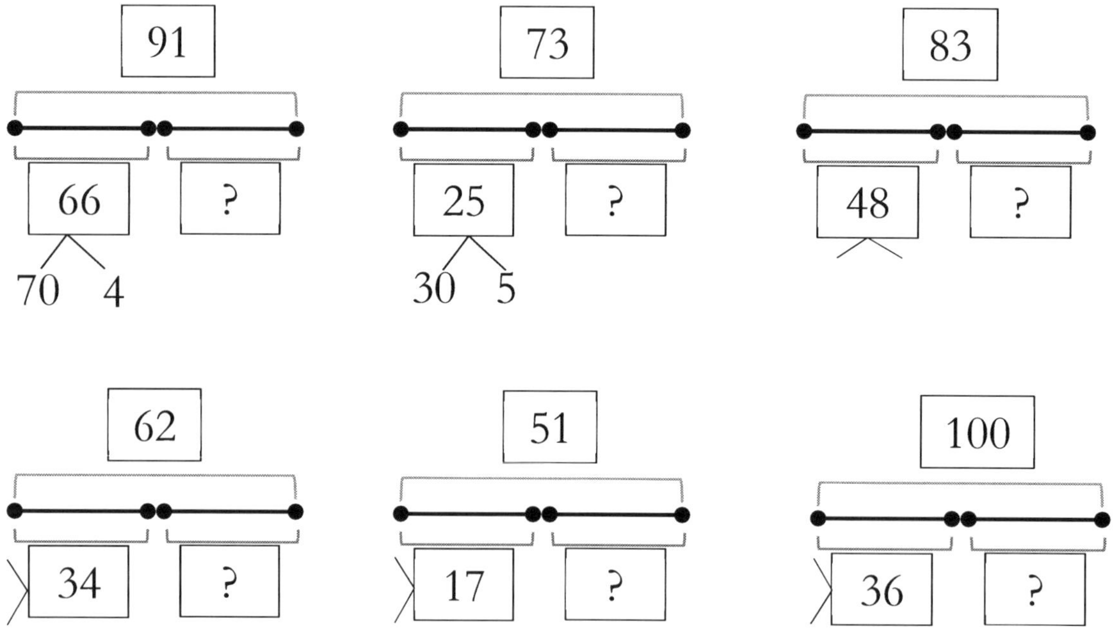

3. Write the missing numbers and an equation. Subtract.

40 - 27 → 30, 3

27 = 30 - 3

40 - 30 + 3 = __

60 - 32 → __, __

__ = __ - __

__ - __ + __ = __

80 - 59 → __, __

____

50 - 16 → __, __

____

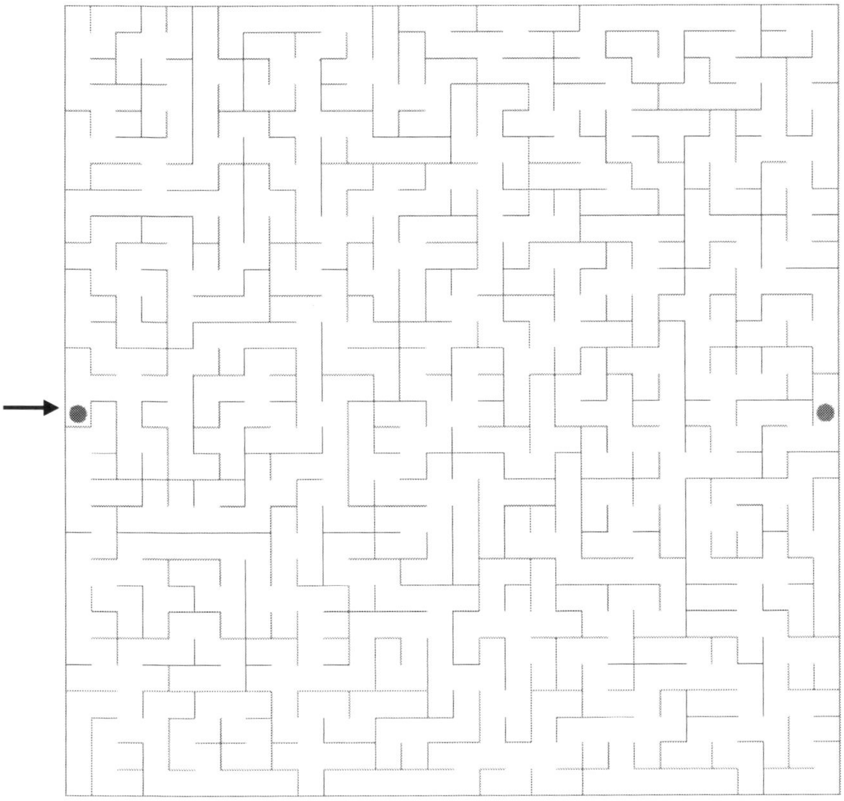

1. Find and circle or cross out the words to find out what you can catch fishing in Minecraft.

S	K	F	S	A	D	D	L	E	S
B	T	F	S	H	H	S	H	P	B
O	P	I	O	K	D	O	S	S	C
W	R	O	C	O	O	S	I	H	B
L	K	T	R	K	E	O	F	B	H
S	U	R	W	L	S	D	B	O	L
O	F	A	T	S	E	N	O	B	H
M	K	T	Y	J	G	P	G	B	R
C	O	P	R	X	K	C	J	P	F
B	D	E	R	S	R	N	Z	D	S

FISH
BOWLS
STICKS
RODS
HOOKS
BOTTLES
BONES
BOOKS
SADDLES

1. Compare and find by how many more or less the numbers are different.

80 > 20 + 20 by 40	100 - 55 < 70 by __
90 - 60 > 10 by __	60 < 36 + 44 by __
20 < 10 + 90 by __	43 + 27 < 100 by __
70 - 50 > 10 by __	30 < 53 + 38 by __
95 > 30 + 50 by __	95 - 47 < 64 by __

2. Subtract. Decompose the subtrahend.

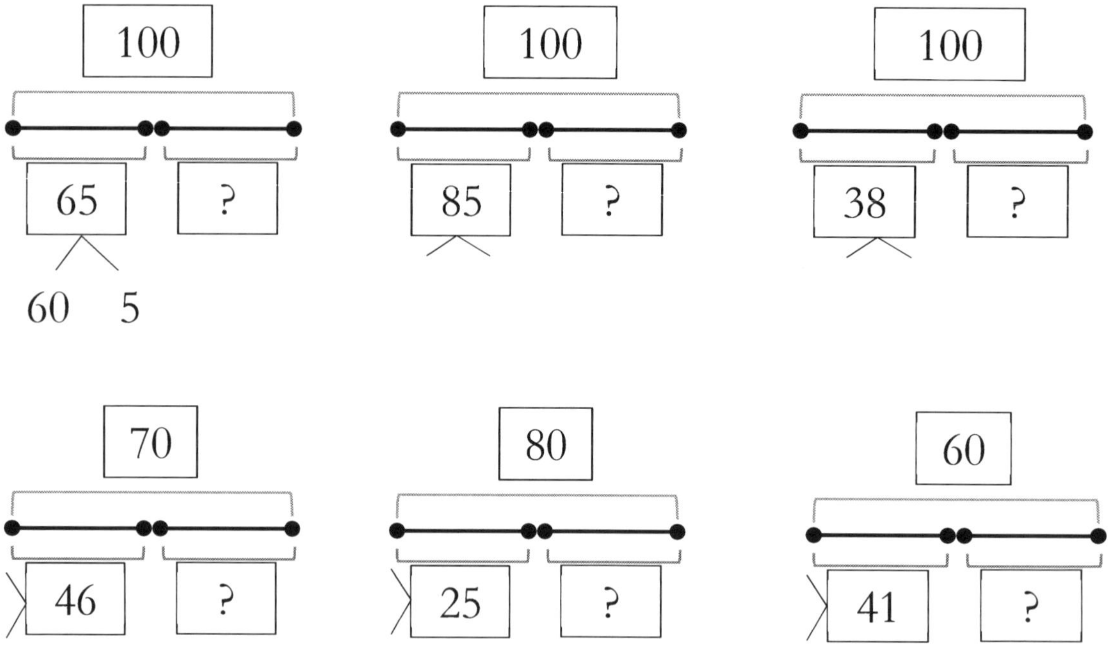

1. Fill in the missing numbers to make the diagrams true.

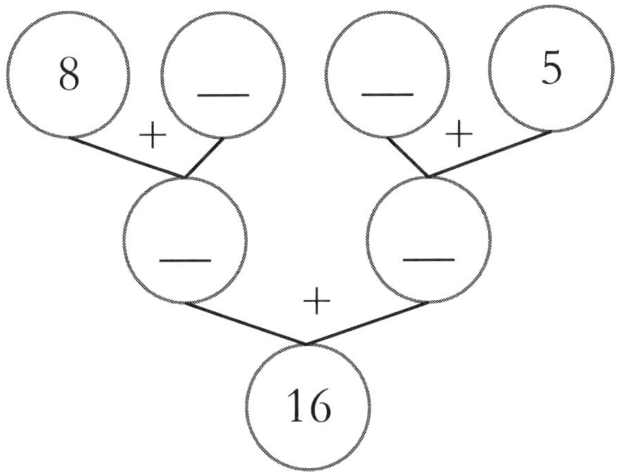

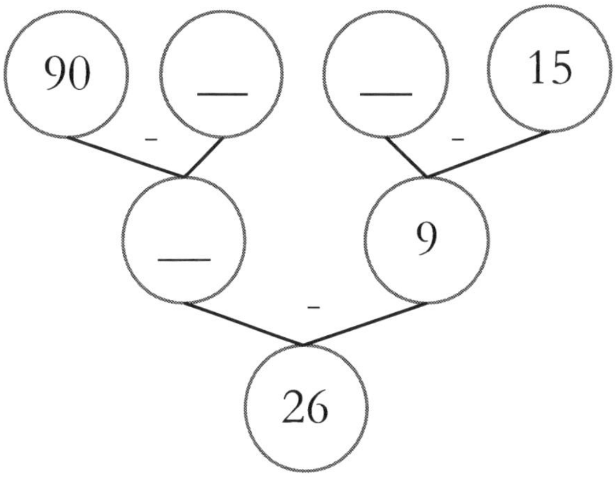

2. Subtract.

142 - 89 = ______________________________________

126 - 75 = ______________________________________

131 - 79 = ______________________________________

3. Fill in the missing numbers (__ equals one digit) to make the equations true.

__ __ + 1 = __ __ __

__ __ __ + 1 = __ __ __ __

__ __ __ __ __ - 1 = __ __ __ __

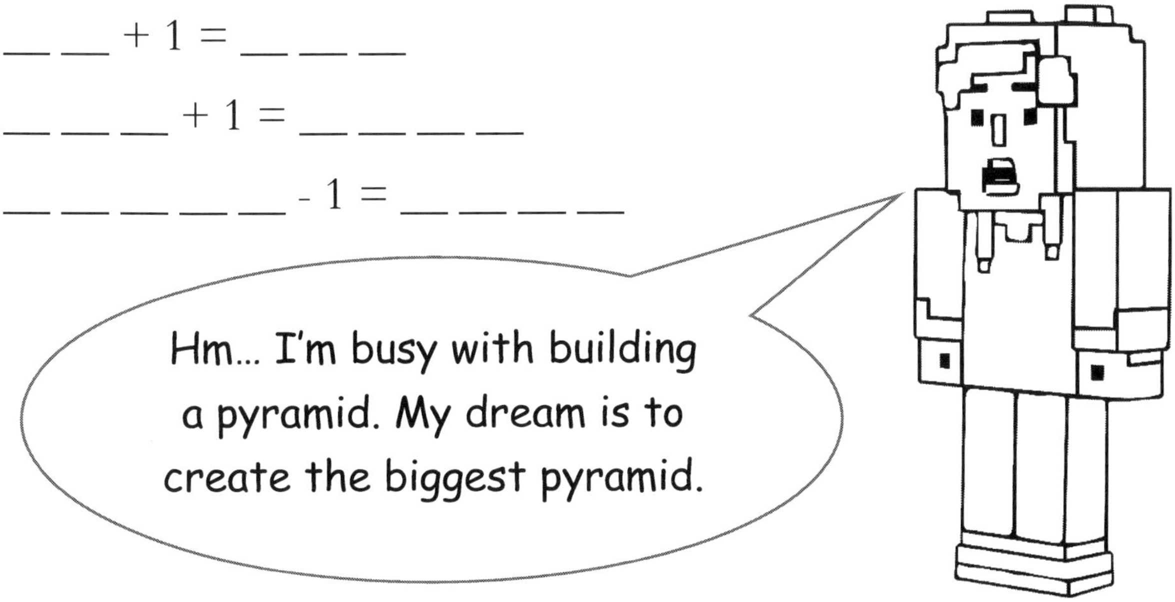

Hint: Borrow from the next figure to the left;
When borrowing 1 ten, change it to 10 ones;
When borrowing 1 hundred, change it to 10 tens.

72 - 38

32 **6**

Decompose 38.
38 is 32+6.

	tens	ones
	7	2
-	3	8
	__	__

In columns we subtract 8 ones from 2 ones.

But you cannot subtract 8 ones from 2 ones: 2 – 8 = ?!? You must get more ones.

Wait… 72 - 32 – 6 = 40 - 6 = 34.

Step 1: If I subtract in columns, I need one more row above 72.

	tens	ones
	__	__
	7	2
-	3	8
	__	__

Step 2: I borrow 1 ten from 7 tens since 70=10+60.

10 borrowed ones + 2 ones are 12 ones, write 12 above 2 in one's place. Write the leftover 6 tens above 7 in ten's place.

	tens	ones
	6	12
	7	2
-	3	8
	__	__

Step 3: Cross out 7 and 2 to avoid mistakes.

Step 4: Subtract 8 ones from 12 ones: 12-8=4.

Step 5: Subtract 3 tens from 6 tens: 6 – 3 = 3.

	tens	ones
	6	12
	~~7~~	~~2~~
-	3	8
	3	4

Hint: Write ones under ones, tens under tens, and so on;
Add ones first, then tens, then hundreds;
Carry when the sum in any column is 10 or more.

44 + 29

6 **23**

Decompose 29.
29 is 6+23.

	tens	ones
	4	4
+	2	9
	__	__

4+9=13. So, you write 3 in one's place and carry 1 ten with the other tens.

Aha, I need to add one more row to write the numbers that were carried, right?

Wait… 44 + 29 = 44 + 6 + 23 = 50 + 23 = 73.

	tens	ones
	__	__
	4	4
+	2	9
	__	__

Step 1: If I add in columns, I need one more row above 44.

Step 2: First, I add ones: 4+9=13.

Write the 3 in one's place. Carry 1 ten with the other tens. Write the 1 above 4.

	tens	ones
	1	__
	4	4
+	2	9
	__	3

Step 3: Add tens: 1+4+2=7
Write the 7 in ten's place.

	tens	ones
	1	__
	4	4
+	2	9
	5	3

1. Subtract. Use the strategy below: 100–76=100–80+4 = 24.

100-50+5		100-		100	
45	?	15	?	64	?

90		50		80	
77	?	25	?	55	?

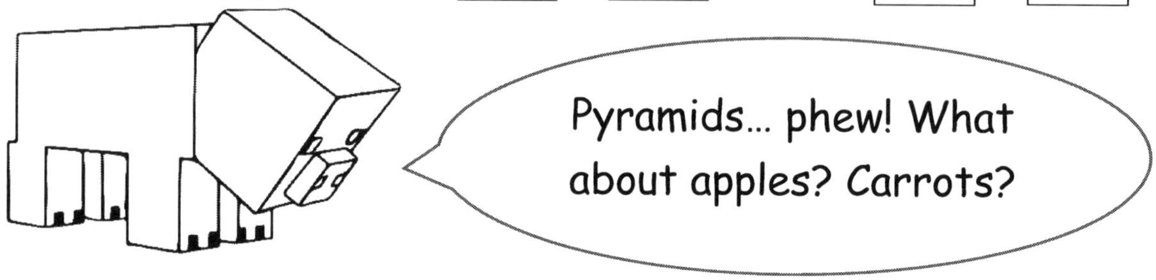

2. Add or subtract.

39 + 13

1 12

tens ones
1
3 9
\+ 1 3
___ ___

39 + 13 = 39 + 1 + 12 =___+___= ___

43 - 24

23 1

tens ones
___ ___
___ ___
\- ___ ___
___ ___

43 – 24 = ___ - ___ - ___ =___- ___= ___

29 + 59

___ ...

tens ones
___
___ ___
\+ ___ ___
___ ___

29 + 59 = ___

____

93 - 36

___ ___

tens ones
___ ___
___ ___
\- ___ ___
___ ___

93 – 36 = ___

____

1. Circle the right answer.

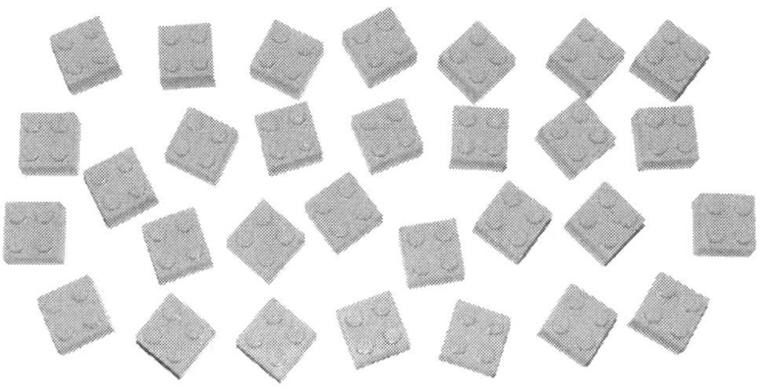

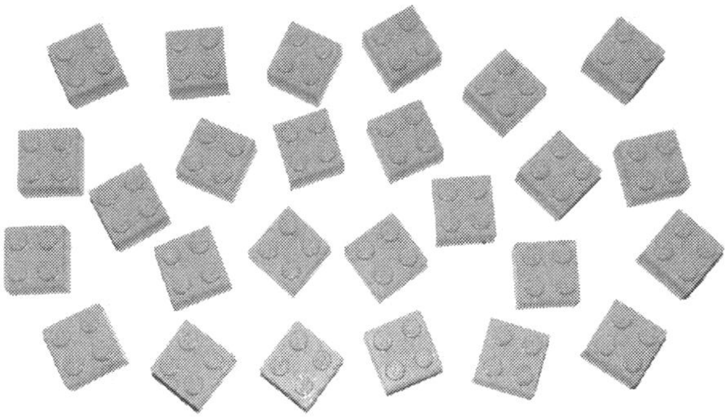

A

B

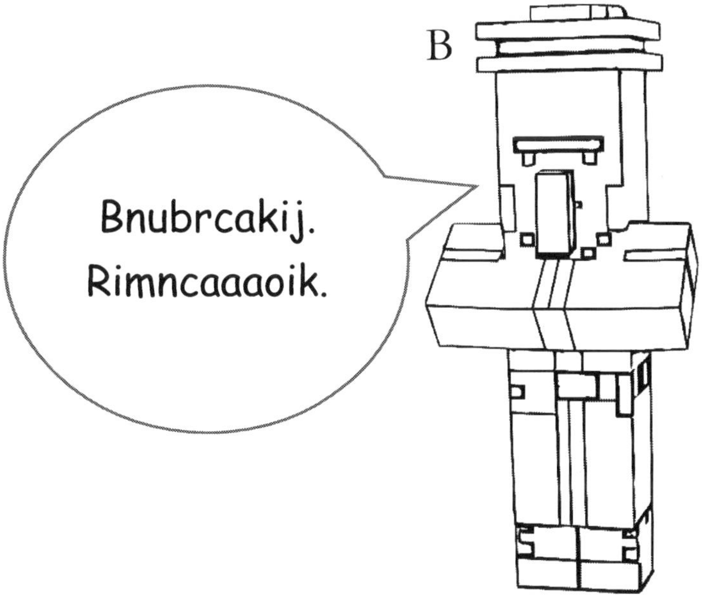

A is greater than B

A is less than B

A is equal to B

2. Subtract. Follow Round-Up-Strategy.

100-20+5	100	100
15 — ?	76 — ?	55 — ?

50	40	70
38 — ?	18 — ?	29 — ?

1. Write the missing numbers and an equation. Add or subtract.

Hint: use Round-Up-Strategy.

46 + 29	51 - 34	68 + 15	94 - 47
30 __	40 __	__ __	__ __
tens ones	tens ones	tens ones	tens ones
__	__ __	__	__ __
__ __	__ __	__ __	__ __
+ __ __	- __ __	+ __ __	- __ __
__ __	__ __	__ __	__ __

46 + 29 = __ + __ - __ = __

51 – 34 = __ - __ + __ = __

68 + 15 = ____ ____________

94 – 47 = ____ ____________

2. Subtract. Hint: use Round-Up-Strategy.

1000 – 95 = __ 1000 – 82 = __ 1000 – 38 = __

800 – 64 = __ 800 – 79 = __ 800 – 41 = __

3. Add or subtract.

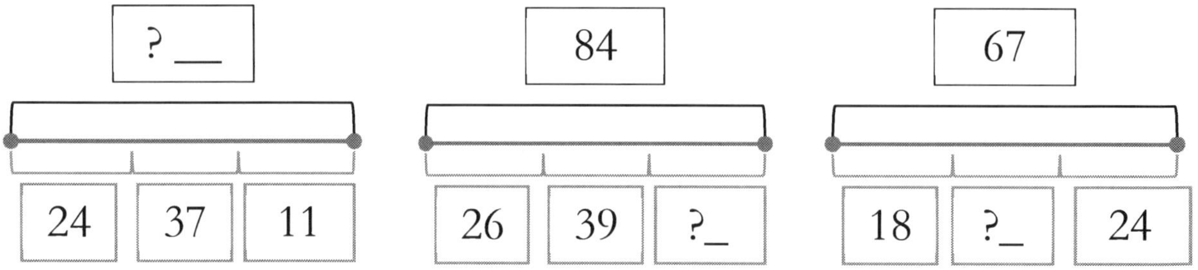

1. Write the missing numbers 15, 23, 26, 27, and 45.

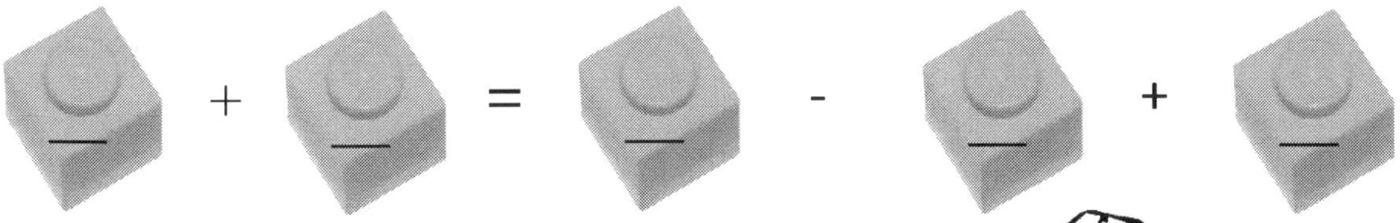

___ + ___ = ___ - ___ + ___

2. Answer the question.

A	B	C	D	E	F	G	H	I	J	K	L	M
1	___	___	___	___	___	___	___	___	___	___	___	___
N	**O**	**P**	**Q**	**R**	**S**	**T**	**U**	**V**	**W**	**X**	**Y**	**Z**
___	___	___	___	___	___	___	___	___	___	___	___	26

_ _ _ _   _ _   _ _ _   _ _ _ _   _ _

23 8 1 20   4 15   25 15 21   14 5 5 4   20 15

_ _ _ _ _   _ _ _ _ _ _ _ _ _?

3 18 1 6 20   19 1 14 4 19 20 15 14 5

____.

1. Add or subtract.

58 + 25	83 - 75	38 + 34	82 - 68
2 __	73 __	__ __	__ __
tens ones	tens ones	tens ones	tens ones
__	__ __	__	__ __
__ __	__ __	__ __	__ __
+ __ __	- __ __	+ __ __	- __ __
__ __	__	__ __	__ __

58 + 25 = ___ 83 – 75 = ___ 38 + 34 = ___ 82 – 68 = ___

__________ __________ __________ __________

2. Fill in the missing numbers ("__" equals one digit).

102 – __ = __ 3 105 - __ = __ 8 111 – __ 3 = 9 __

3. The sum of 2 numbers is 40 and their difference is 16. Find these numbers:________________________________.

4. Continue a series of numbers:

5, 6, 8, 9, 11, _____, _____, _____, _____, _____, _____

1. Fill in the missing whole numbers.

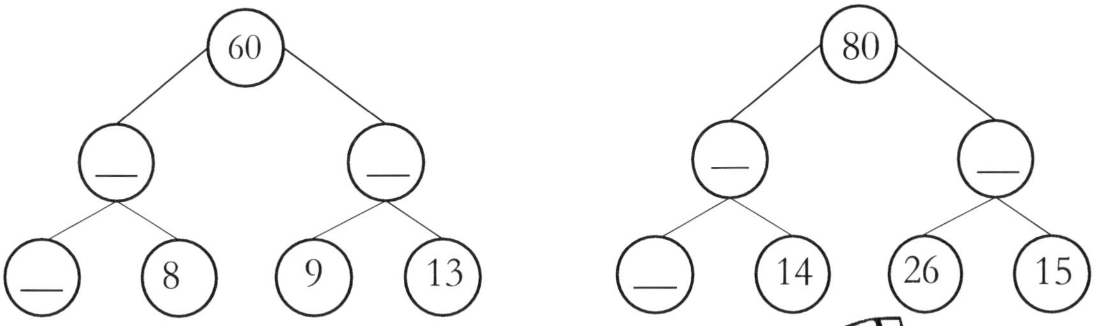

2. Unscramble each word to find out what flowers you can generate in Minecraft. Write the word in the squares.

chrodi

eors

cilla

noildeand

wrensuflo

Guys, what is Rounding?!

I LOVE rounding! Rounding is a simpler and shorter way to represent a number. But you need to keep an approximately equal value.

Eh?!.. Do you hear THAT?! Do you have a fever?

Listen, I have $24. I need to round it to the nearest ten. I explore the number range from 20 to 30:

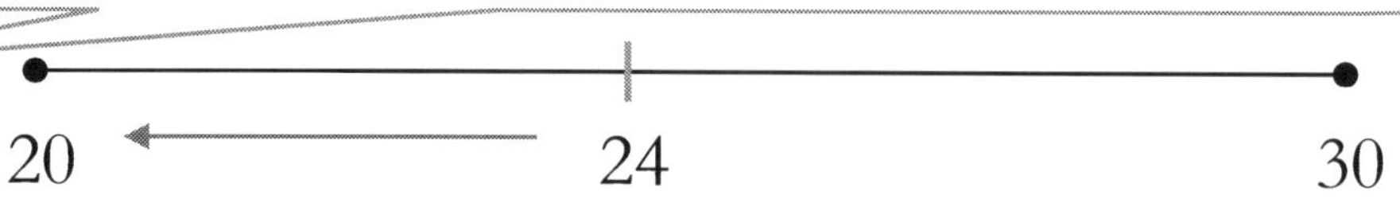

24 is close to 20 since $24 - 20 = 4$

24 is not as close to 30 since $30 - 24 = 6$

$4 < 6$, so, 24 is closer to 20 than to 30.

If you round a number followed by 1,2,3, and 4, you round the number down: 24 is 20 or 62 is 60.

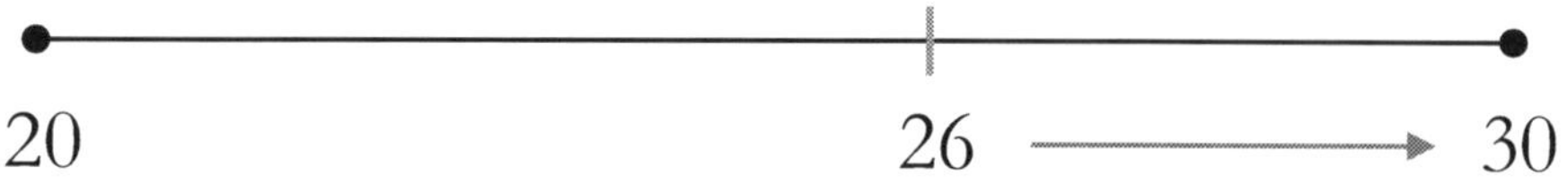

I have $26. I need to round it to the nearest ten.

26 is close to 30 since $30 - 26 = 4$

26 is not as close to 20 since $26 - 20 = 6$

$6 > 4$, so, 26 is closer to 30 than to 20.

If you round a number followed by 6,7,8, and 9, you round the number up: 26 is 30 or 87 is 90.

Ha! Looks like these numbers don't like to run☺.

Look, the numbers always run to the distance that is less. The number 24 runs to 20 but not to 30, right?

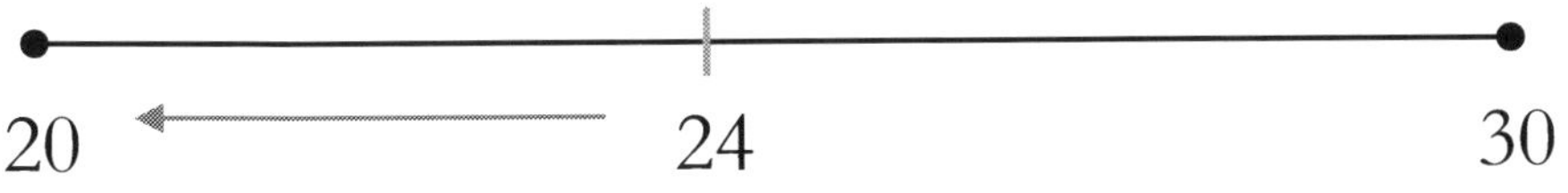

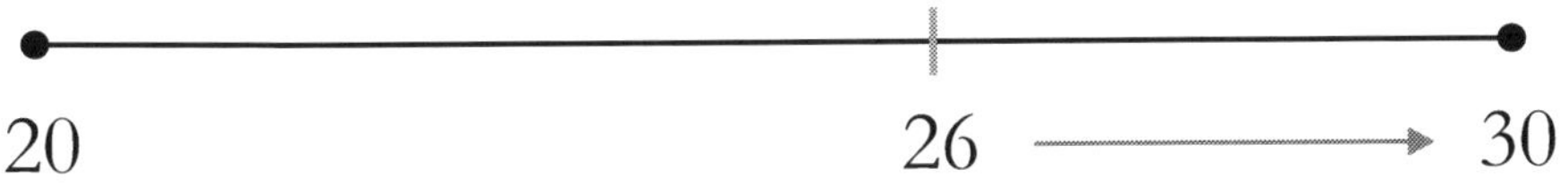

But <u>how</u> can I round $25?

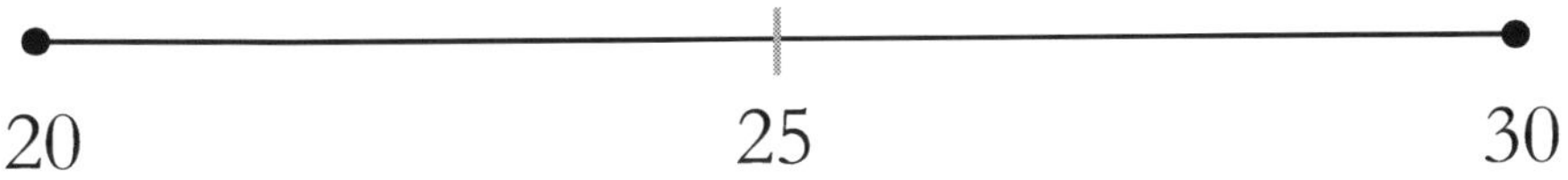

25 is exactly between 20 and 30:

25 is close to 30 since 30 – 25 = 5

25 is also close to 20 since 25 – 20 = 5

25 is Lucky! The numbers ending in 5 always round up! 5 rounds to 10, 15 to 20, 25 to 30, 35 to 40, 45 to 50, 55 to 60, 65 to 70, 75 to 80, 85 to 90, 95 to 100. Done!

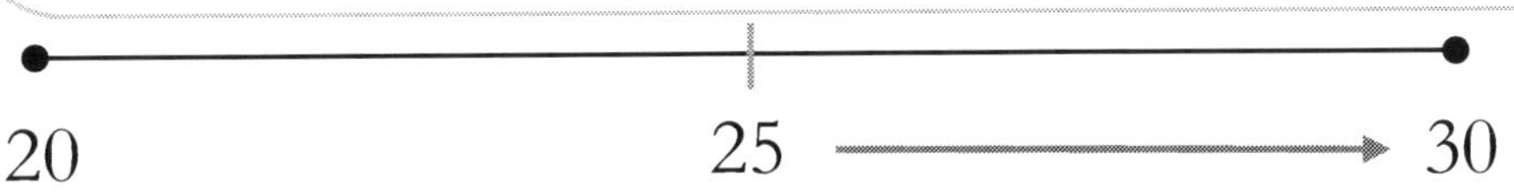

1. Round the numbers to the nearest ten.

89 is ______   63 is ______   47 is ______

22 is ______   34 is ______   75 is ______

91 is ______   56 is ______   18 is ______

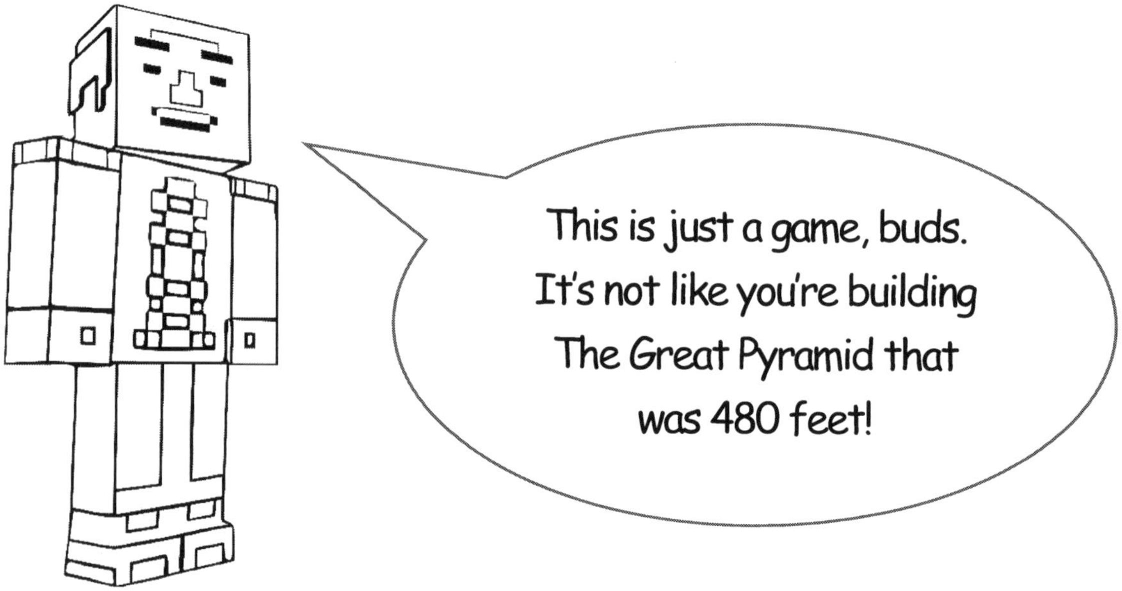

2. Add or subtract.

46 + 45	73 - 27	68 + 16	36 - 18
4 __	23 __	20 __	20 __
tens ones	tens ones	tens ones	tens ones

46 + 45 =   73 – 27 =   68 + 16 =   36 – 18 =

1. Do you like a challenge? I DO! Let's take a timer and see who will solve these problems faster. Adults are welcome, too!

Name ______________________________ Time _____ : _____.

9 + 2 = __	11 – 2 = __	8 + 3 = __	11 – 3 = __
9 + 3 = __	12 – 3 = __	8 + 4 = __	12 – 4 = __
9 + 4 = __	13 – 4 = __	8 + 5 = __	13 – 5 = __
9 + 5 = __	14 – 5 = __	8 + 6 = __	14 – 6 = __
9 + 6 = __	15 – 6 = __	8 + 7 = __	15 – 7 = __
9 + 7 = __	16 – 7 = __	8 + 8 = __	16 – 8 = __
9 + 8 = __	17 – 8 = __	8 + 9 = __	17 – 9 = __
9 + 9 = __	18 – 9 = __	6 + 5 = __	11 – 5 = __
5 + 6 = __	11 – 6 = __	6 + 6 = __	12 – 6 = __
5 + 7 = __	12 – 7 = __	6 + 7 = __	13 – 7 = __
5 + 8 = __	13 – 8 = __	6 + 8 = __	14 – 8 = __
5 + 9 = __	14 – 9 = __	6 + 9 = __	15 – 9 = __
7 + 4 = __	11 – 4 = __	4 + 7 = __	11 – 7 = __
7 + 5 = __	12 – 5 = __	4 + 8 = __	12 – 8 = __
7 + 6 = __	13 – 6 = __	4 + 9 = __	13 – 9 = __
7 + 7 = __	14 – 7 = __	3 + 8 = __	11 – 8 = __
7 + 8 = __	15 – 8 = __	3 + 9 = __	11 – 9 = __
7 + 9 = __	16 – 9 = __	8 + 5 = __	13 – 5 = __

1. Round the numbers to the nearest ten.

895 is ______	634 is ______	476 is ______
222 is ______	349 is ______	753 is ______
917 is ______	562 is ______	188 is ______
146 is ______	328 is ______	856 is ______
925 is ______	751 is ______	117 is ______
367 is ______	423 is ______	191 is ______

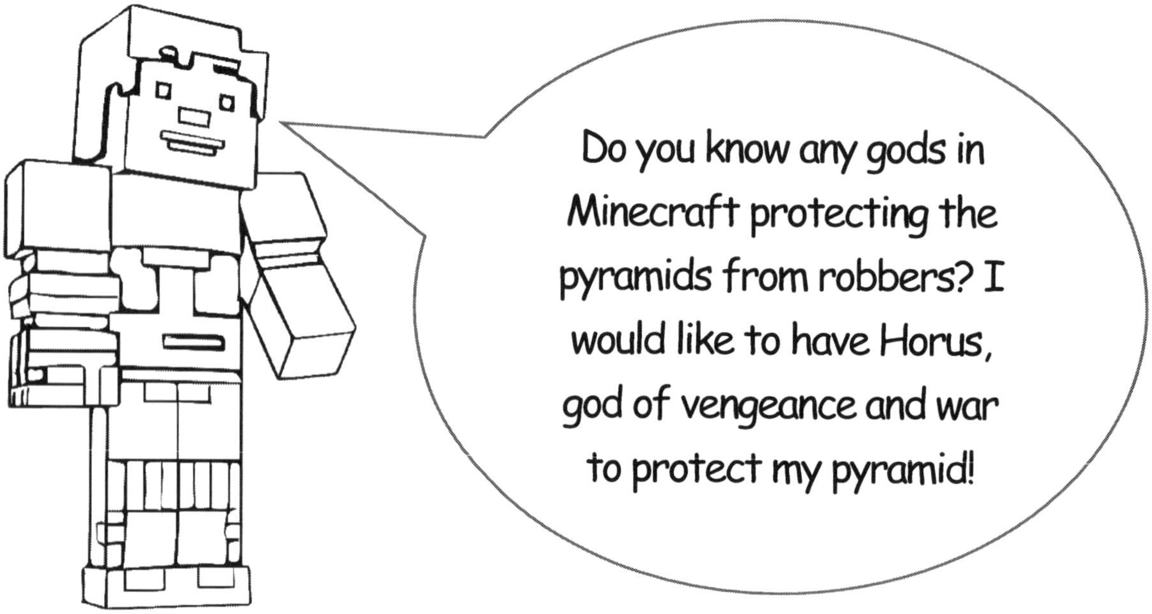

2. Write the missing numbers and an equation. Subtract.

44 - 29
/\
30   1

29 = 30 - 1
44 - 30 + 1 = __

37 - 19
/\
__   __

__ = __ - __
__ - __ + __ = __

61 - 35
/\
__   __

____

83 - 46
/\
__   __

____

1. Find the missing number.

100 - X = 67	100 - X = 28	100 - X = 2
X = ______	X = ______	X = ______
100 - ? = 46	100 - ? = 24	100 - ? = 71
? = ______	? = ______	? = ______
100 - ? = 45	100 - ? = 73	100 - ? = 39
? = ______	? = ______	? = ______

2. Round the numbers to the nearest hundred.

140 is _____	370 is _____	590 is _____
220 is _____	860 is _____	630 is _____
910 is _____	480 is _____	750 is _____

3. Find the value. Draw the arrows to make the tens.

3 + 1 + 7 + 6 + 4 + 9 + 4 + 7 + 8 + 12 + 16 + 13 = __________

13 + 11 + 8 + 5 + 4 + 9 + 15 + 7 + 12 + 16 = __________

12 + 15 + 11 + 14 + 17 + 19 + 18 + 13 + 15 + 16 = __________

1. Fill in the missing numbers. A number has to be more than 1.

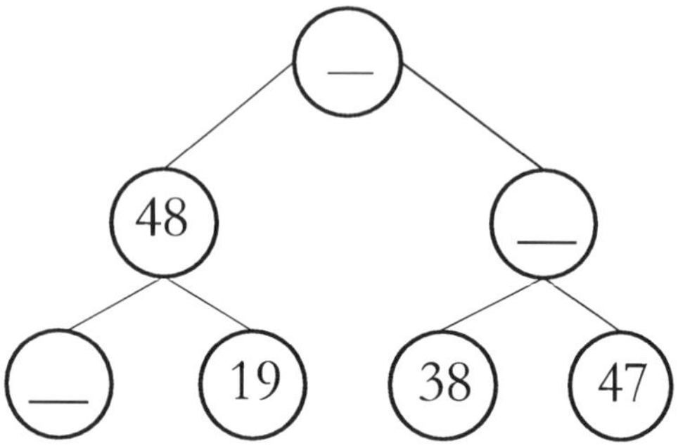

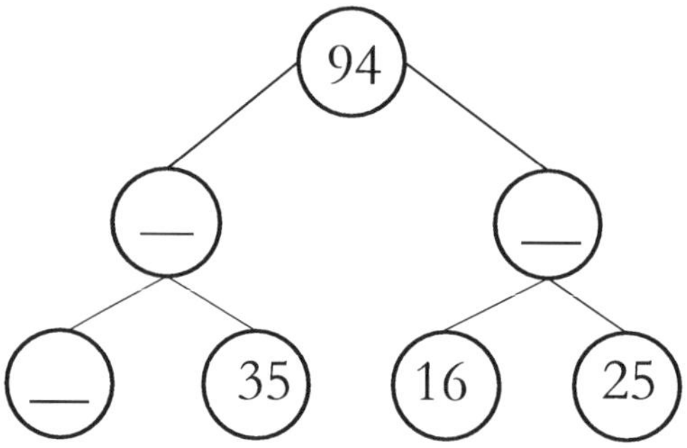

2. Find and circle or cross out the words to find out more about horses in Minecraft.

M	F	M	W	S	A	Q	N	R	B
S	C	T	O	Q	S	I	Z	E	R
G	Q	V	N	Y	K	P	Y	B	E
R	A	G	U	S	Q	R	E	W	E
D	B	P	K	O	T	L	U	E	D
R	C	C	L	N	E	N	G	L	D
L	U	G	X	A	W	K	U	A	E
B	W	Z	S	K	P	P	B	O	R
I	X	H	O	V	C	D	V	F	M
A	P	P	L	E	S	R	C	S	L

SPEED
FOAL
BREED
LEASH
APPLES
SUGAR
MOUNT
BUCKSKIN

1. Compare and find by how many more or less the numbers are different.

68 >	33	by	35	27	95	by	__
14	51	by	__	80	45	by	__
65	17	by	__	28	77	by	__
66	96	by	__	91	18	by	__

2. Write the missing numbers near the letters to make the equations true.

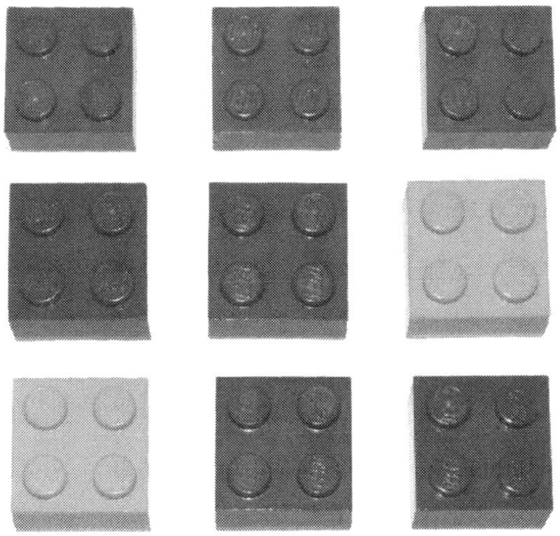

G __ + G __ + G __ = 24

G __ + G __ + Y __ = 31

Y __ - B __ + G __ = 22

3. Find the value.

739586: tens + ten thousands – ones + hundreds + hundred thousands – thousands = ______________________.

Let's go through rounding one more time, eh? I don't get it. Lets' take 242 and round it to the nearest hundred. 242 may be rounded up to 300 or down to 200. Which one is correct?

Let's look at the range from 200 to 300. I will write 2 and 3 to show 2 hundreds (200) and 3 hundreds (300)

I write the numbers 0, 1, 2, 3, 4, 5, 6, 7, 8, 9 to show tens.

2 0 1 2 3 4 5 6 7 8 9 3

200 ← → 300

There are 10 digits and we divide them equally into 2 parts: 0, 1, 2, 3, 4. Then, we separate them from 5, 6, 7, 8, 9.

0, 1, 2, 3, 4 are close to 2 hundreds. So, when rounding, we go down and the numbers 208, 215, 227, 239, 246 are rounded down to 200 (look only at the next digit to the right – I underline the tens if I need to round to the nearest hundred).

5, 6, 7, 8, 9 are close to 3 hundreds.

So, when rounding, we go up and the numbers 259, 266, 278, 284, 297 are rounded up to 300.

1. Write the missing numbers and an equation. Add. Use Round-Up-Strategy.

39 + 26	74 + 17	48 + 39
30 __	__ __	__ __
tens ones	tens ones	tens ones
__	__	__
__ __	__ __	__ __
+ __ __	+ __ __	+ __ __
__ __	__ __	__ __

39 + 26 = ______ 74 + 17 = ______ 48 + 39 = ______

______ ______ ______

2. Subtract.

64	93	58
− 4: 28, ?__	− 3: 66, ?__	− 1: 29, ?__

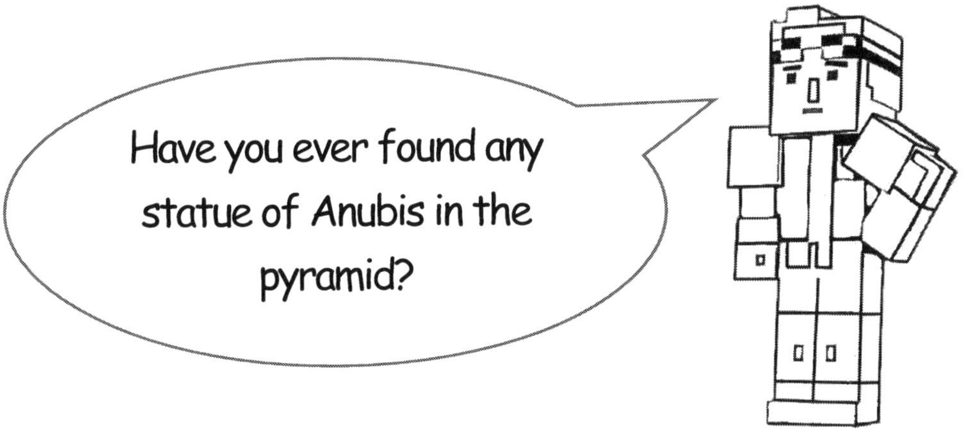

1. Fill in the missing "+", "-", or "=" to make the equations true. You can use the same sign ("+", "-", or "=") more than once.

12 _____ 31 _____ 43 _____ 62

96 _____ 85 _____ 18 _____ 29

25 _____ 32 _____ 21 _____ 36

0 _____ 96 _____ 64 _____ 32

76 _____ 21 _____ 42 _____ 55

28 _____ 81 _____ 65 _____ 12

1. Subtract. Hint: 500 – 290 = 500 – 200 – 90 = 210.

800 – 260 = __ 400 – 170 = __ 300 – 150 = __

900 – 530 = __ 600 – 280 = __ 700 – 640 = __

2. Write the number words.

509 ____

____

3. Subtract.

775 - 525 = __ 750 - 325 = __ 925 - 575 = __

975 - 425 = __ 450 - 125 = __ 625 - 475 = __

4. Round the numbers to the nearest hundred.

845 is ______ 684 is ______ 876 is ______

226 is ______ 549 is ______ 723 is ______

987 is ______ 669 is ______ 184 is ______

116 is ______ 358 is ______ 356 is ______

924 is ______ 721 is ______ 157 is ______

467 is ______ 473 is ______ 491 is ______

1. Write the missing numbers and an equation. Add or subtract.

Use Round-Up-Strategy.

68 + 14	82 - 37	59 + 16	72 - 36
20 __	40 __	__ __	__ __
tens ones	tens ones	tens ones	tens ones
__	__ __	__	__ __
__ __	__ __	__ __	__ __
+ __ __	- __ __	+ __ __	- __ __
__ __	__ __	__ __	__ __

68+14=__+__ -__=__-__=__

82–37=__-__ +__=__+__=__

59 + 16 =

72 – 36 =

2. Subtract.

720 - 280 = __

910 - 430 = __

850 - 290 = __

630 - 270 = __

740 - 80 = __

440 - 380 = __

750 - 150 - 250 = ______

900 - 150 - 350 = ______

800 - 250 - 300 = ______

700 - 250 - 300 = ______

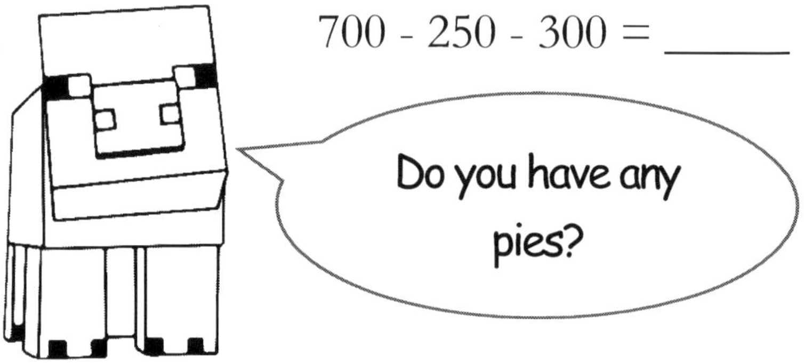

1. Write the missing numbers and an equation. Subtract or add.

48 + 23	65 - 49	57 + 18	84 - 45
2 21	45 4	__ __	__ __
tens ones	tens ones	tens ones	tens ones
__	__ __	__	__ __
__ __	__ __	__ __	__ __
+ __ __	- __ __	+ __ __	- __ __
__ __	__ __	__ __	__ __

48 + 23 =
__+__+__ =
__+__ =__

65 – 49 = __-
__-__ = __-
__=__

57+18 =

84–45 =

2. Fill in the missing numbers to make the equations true.

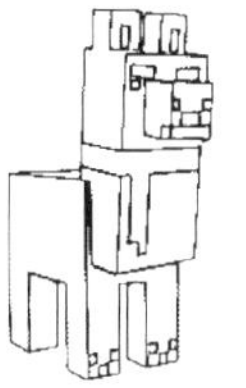
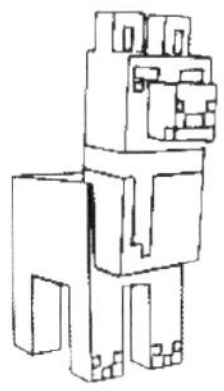
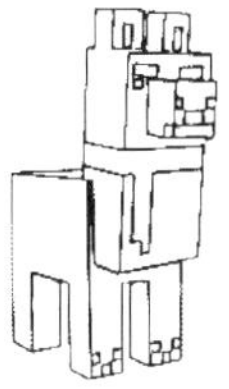

__ + __ + __ = 45

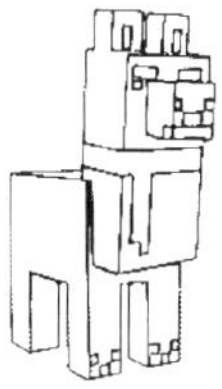
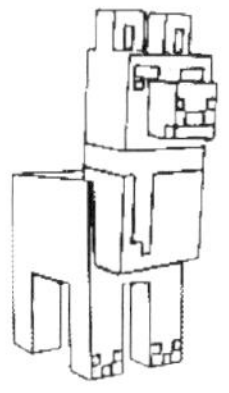
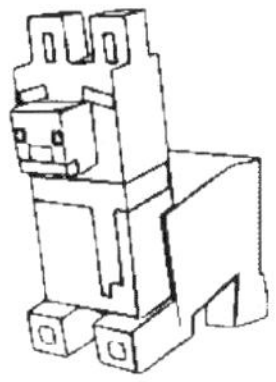

__ + __ + __ = 39

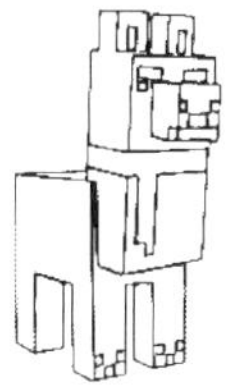
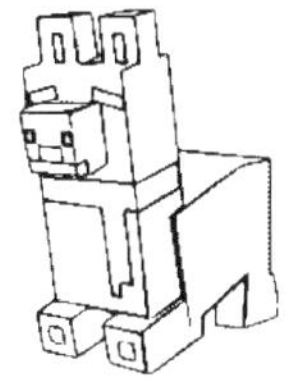

__ + __ + __ = 32

1. I have a favorite two-digit number:

- the sum of tens (T) and ones (O) is 15;
- their difference is 3;
- the ones digit is an even number.

What number is my favorite? __ __.

T + O = __ T - O = __ O = ______ __ __?

2. Continue a series of numbers.

3, 6, 12, 15, 21, _____, _____, _____, _____, _____, _____.

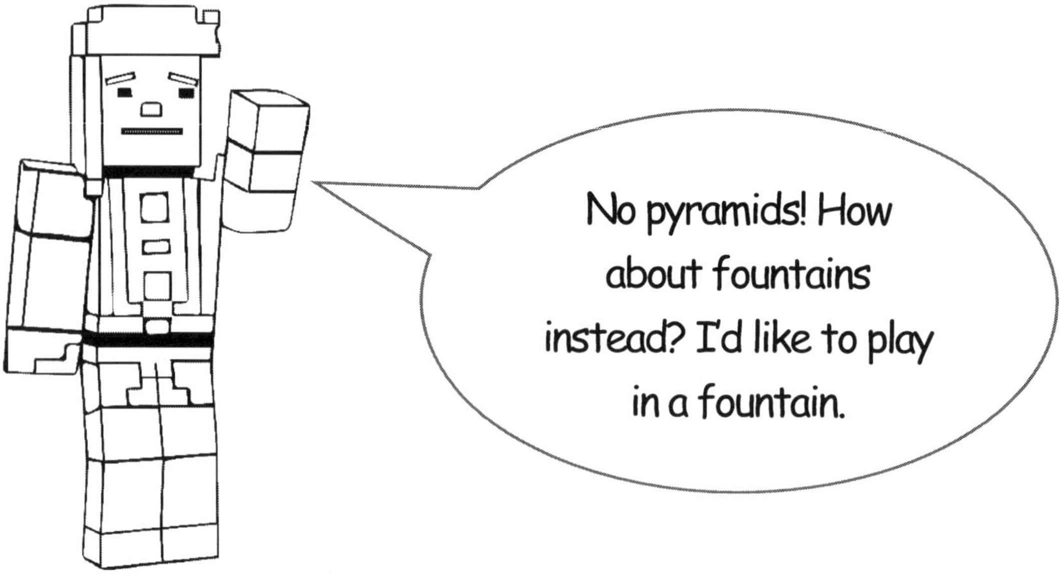

3. Fill in the missing signs "+" or/and "-" to make the equations true.

5 5 5 5 5 = 25 5 5 5 5 5 = 5

5 5 5 5 5 = 15 5 5 5 5 5 5 = 20

1. Write the missing numbers and an equation. Subtract.

76 - 28	96 - 49	57 - 38
26 __	46 __	__ __
tens ones	tens ones	tens ones
__ __	__ __	__ __
__ __	__ __	__ __
- __ __	- __ __	- __ __
__ __	__ __	__ __

76 - 28 = __ - __ - __ = __ - __ = __

96 - 9 = ________
____

57 - 38 = _______
____

2. Subtract.

546 - 108 = ____

630 - 128 = ____

180 - 123 = ____

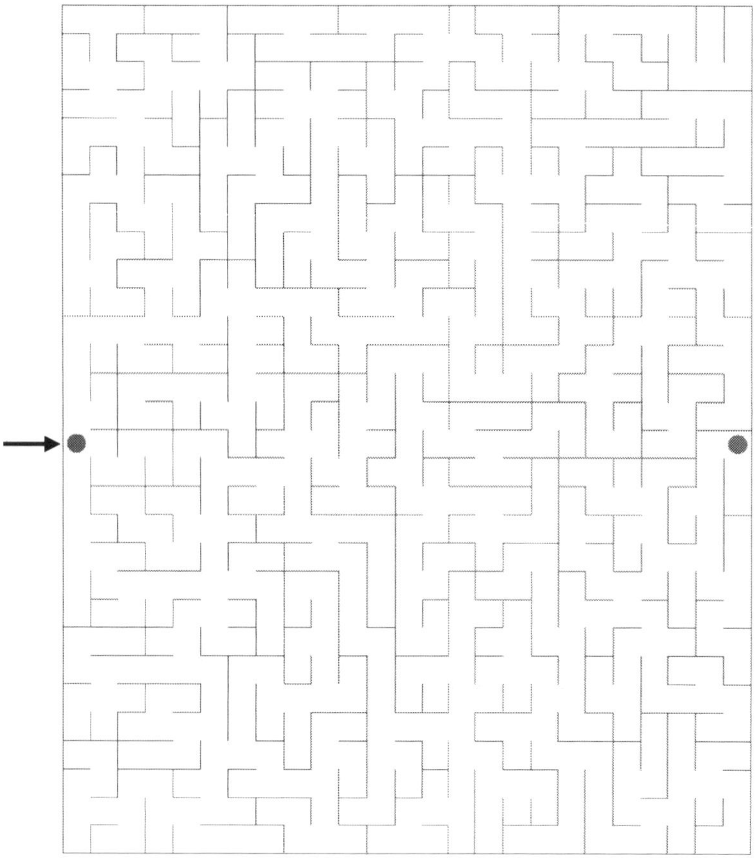

1. Unscramble each word to find the color of the horses in Minecraft . Write the word in the squares.

tsechunt

hewti

calkb

agyr

1. Fill in the missing numbers so that the sums in the columns, rows, and two numeral diagonals are the same. Draw the arrows from the least number to the next greater number, and then to the greatest. Do you see any pattern? ________________.

56	21	28
___	___	___
42	___	14

___	28	___
___	___	4
16	12	32

2. Subtract.

450 - 162 = ____ 975 - 338 = ____ 715 - 242 = ____

493 - 364 = ____ 526 - 370 = ____ 763 - 274 = ____

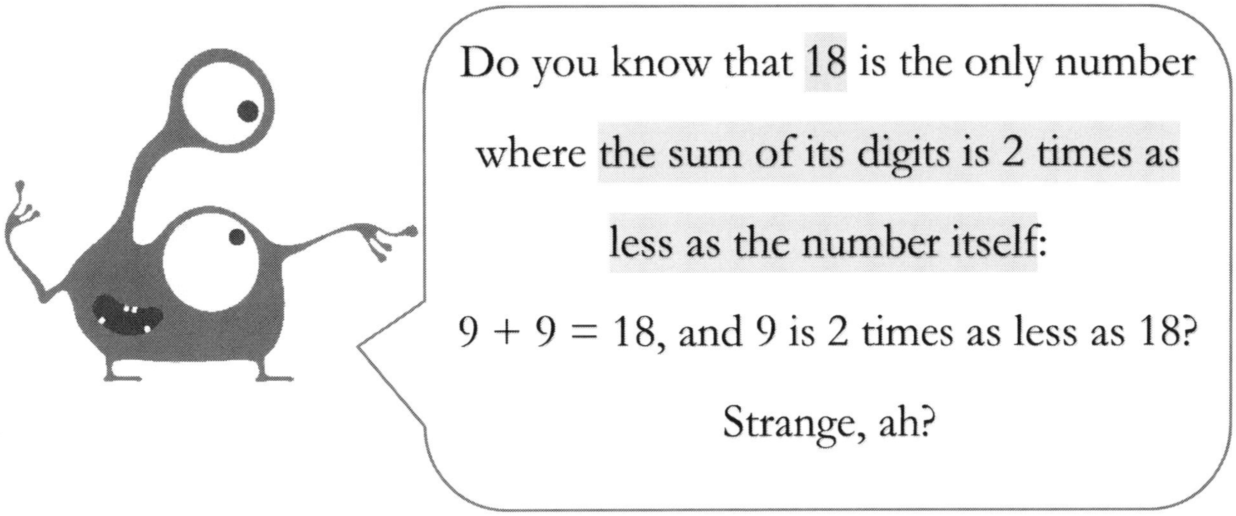

1. Write the missing numbers and an equation. Subtract or add. Follow the strategy.

58 + 25	22 - 13	17 + 16	34 - 26
2 __	12 __	__ __	__ __
tens ones	tens ones	tens ones	tens ones
__	__ __	__	__ __
__ __	__ __	__ __	__ __
+ __ __	- __ __	+ __ __	- __ __
__ __	__ __	__ __	__ __

58 + 25 = __+__+__= __+__=__

22 – 13 = __- __-__ = __- __=__

17 + 16 =

34 – 26 =

4. Round the numbers to the nearest thousand.

2845 is ______	6184 is ______	8756 is ______
2526 is ______	5549 is ______	2723 is ______
9187 is ______	9669 is ______	1184 is ______
1716 is ______	3528 is ______	3256 is ______
9254 is ______	7121 is ______	1657 is ______
4607 is ______	4873 is ______	4951 is ______

1. Write the missing numbers and an equation. Subtract or add. Use Round-Up-Strategy or decompose.

76 + 18	61 - 36	69 + 25	52 - 29
20 __	40 __	1 __	22 __
tens ones	tens ones	tens ones	tens ones
__	__ __	__	__ __
__ __	__ __	__ __	__ __
+ __ __	- __ __	+ __ __	- __ __
__ __	__ __	__ __	__ __

76 + 18 = __+__-__= __-__=__

61 – 36 = __- __+__= __+__=__

69 + 25 = ___ ___________

52 – 29 = ___ ___________

2. Round the numbers to the nearest ten.

2845 is ______	6184 is ______	8756 is ______
2526 is ______	5549 is ______	2723 is ______
9187 is ______	9669 is ______	1184 is ______
1716 is ______	3528 is ______	3256 is ______
9254 is ______	7121 is ______	1657 is ______
4607 is ______	4873 is ______	4951 is ______

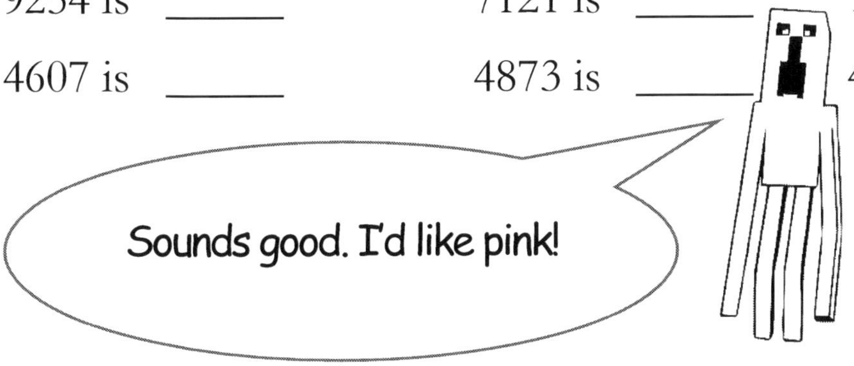

1. <u>Write</u> the missing numbers and an equation. <u>Subtract or add</u>. <u>Follow</u> the strategy.

74 + 19	71 - 35	47 + 35	56 - 27
6 ___	___ 4	40 ___	30 ___
tens ones	tens ones	tens ones	tens ones
___	___ ___	___	___ ___
___ ___	___ ___	___ ___	___ ___
+ ___ ___	- ___ ___	+ ___ ___	- ___ ___
___ ___	___ ___	___ ___	___ ___
74 + 19 = ____	71 – 65 = ____	47 + 35 = ___	56 – 37 = ___
__________	__________	__________	__________

2. <u>Subtract and add</u>. <u>Use</u> Round-Up-Strategy or <u>decompose</u>.

546 - 108 = ___ 630 - 128 = ___ 180 - 123 = ___

840 - 245 = ___ 562 - 158 = ___ 870 - 124 = ___

987 - 351 = ___ 275 - 269 = ___ 753 - 225 = ___

550 + 80 = ___ 273 + 40 = ___ 475 + 50 = ___

161 + 70 = ___ 510 + 85 = ___ 920 + 94 = ___

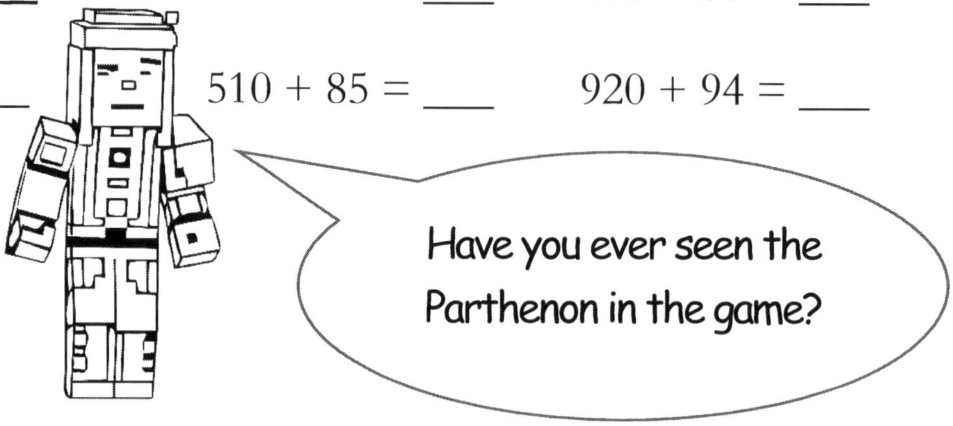

1. Subtract.

64	37	53	95	88
- 31	- 15	- 41	- 74	- 27
__ __	__ __	__ __	__ __	__ __

80	40	60	70	30
- 6	- 7	- 2	- 5	- 9
__ __	__ __	__ __	__ __	__ __

80	40	70	60	30
- 36	- 24	- 52	- 15	- 25
__ __	__ __	__ __	__ __	__ __

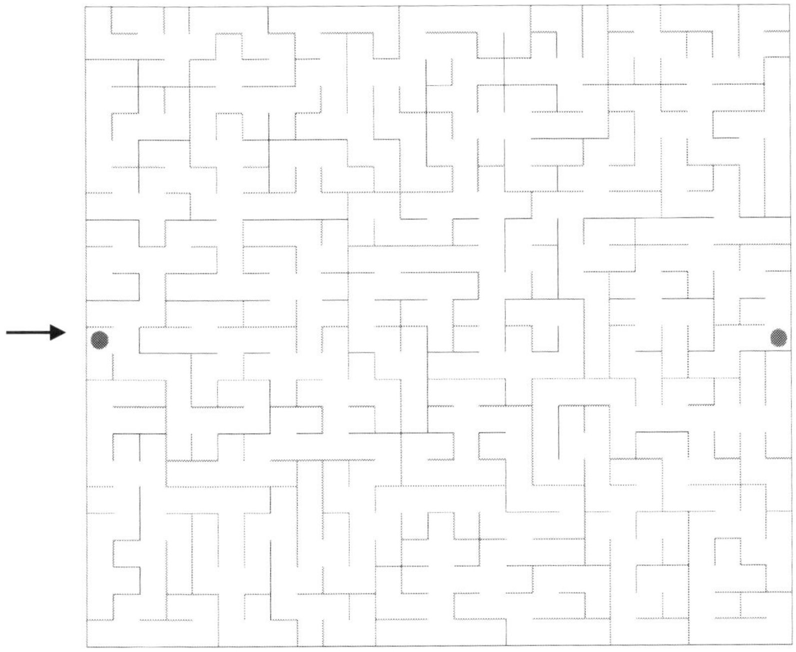

1. Write the missing numbers 12, 16, 18, 31, and 45 to make the equation true.

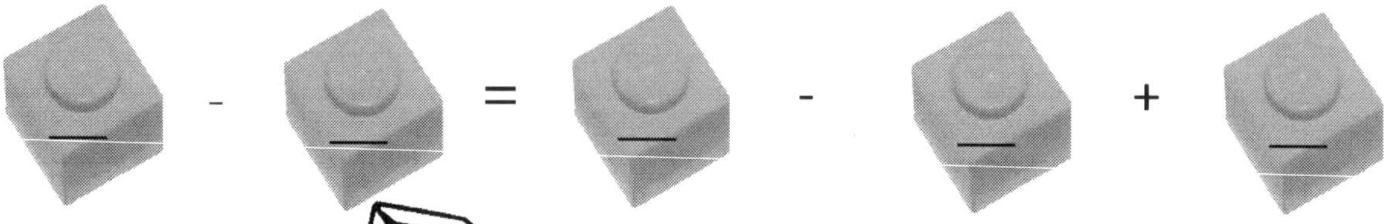

___ - ___ = ___ - ___ + ___

2. Answer the question.

A	B	C	D	E	F	G	H	I	J	K	L	M
1	2	3	4	5	6	7	8	9	10	11	12	13
N	O	P	Q	R	S	T	U	V	W	X	Y	Z
14	15	16	17	18	19	20	21	22	23	24	25	26

___ ___ ___   ___ ___   ___ ___ ___   ___ ___ ___ ___ ___ ___ ___
23 8 15   9 19   20 8 5   12 1 18 7 5 19 20

___ ___ ___   ___ ___   ___ ___ ___   ___ ___ ___ ___?
13 15 2   9 14   20 8 5   7 1 13 5

___ ___ ___   ___ ___ ___ ___ ___   ___ ___ ___ ___ ___ ___
20 18 5   5 14 4 5 18   4 18 1 7 15 14

1. Write the missing numbers and an equation. Subtract or add.

60 - 29	50 - 34	70 - 46
30 ___	40 ___	___ ___
tens ones	tens ones	tens ones
___ ___	___ ___	___ ___
___ ___	___ ___	___ ___
- ___ ___	- ___ ___	- ___ ___
___ ___	___ ___	___ ___

60 – 29 = ______ 50 – 34 = ______ 70 – 46 = ______

______________ ______________ ______________

2. Add.

300 + 400 + 150 = ___ 350 + 450 + 100 = ___

750 + 150 + 100 = ___ 650 + 175 + 125 = ___

550 + 150 + 300 = ___ 450 + 250 + 300 = ___

3. Subtract. Use Round-Up-Strategy.

850 – 80 = __ 250 – 90 = __ 585 – 90 = __

974 – 80 = __ 403 – 50 = __ 109 – 50 = __

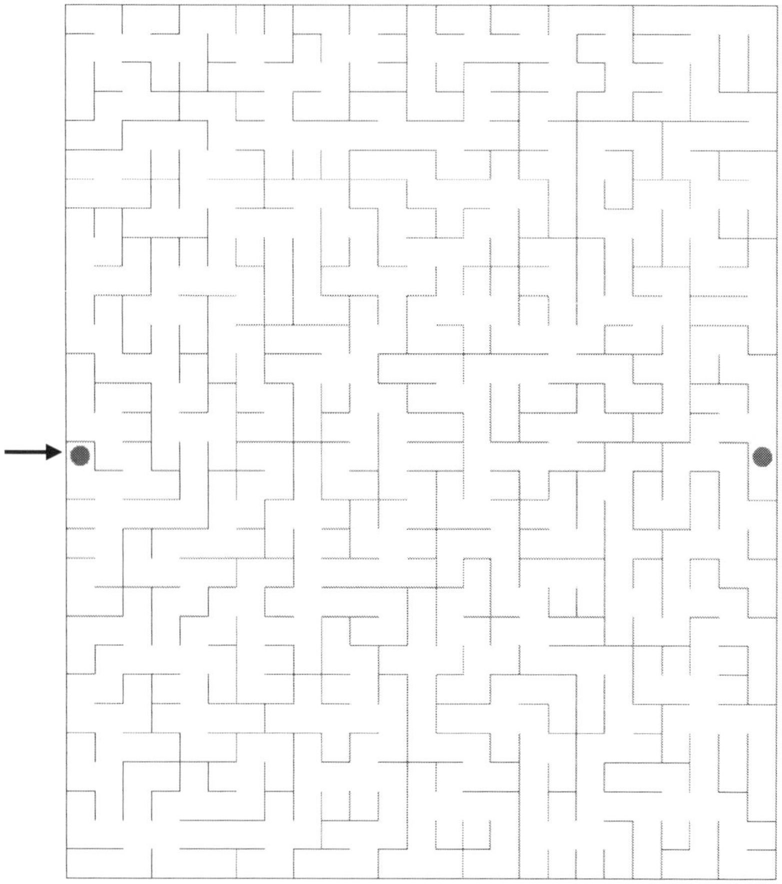

2. <u>Find</u> and <u>circle</u> or <u>cross out</u> the words to find more about lighting in Minecraft.

L	P	T	A	T	R	Y	T	Z	B
I	X	F	H	F	O	R	U	R	Z
G	W	K	N	G	X	R	I	Y	K
H	T	E	R	N	I	G	C	Q	E
T	S	T	E	S	H	L	C	H	T
G	Q	H	T	T	U	V	N	E	B
S	S	E	N	K	R	A	D	U	C
X	A	E	A	M	L	Q	Y	X	S
Y	S	U	L	K	R	L	L	N	H
S	F	U	R	N	A	C	E	S	J

DARKNESS
SUNLIGHT
TORCH
LANTERN
BRIGHTNESS
FURNACE
LIGHT

1. Write the missing numbers and an equation. Subtract or add.

59 + 35	62 - 14	66 + 28	94 - 68
1 __	12 __	__ __	__ __
tens ones	tens ones	tens ones	tens ones
__	__ __	__	__ __
__ __	__ __	__ __	__ __
+ __ __	- __ __	+ __ __	- __ __
__ __	__ __	__ __	__ __

59 + 35 = ___ 62 – 14 = ___ 66 + 28 = ___ 94 – 68 = ___

___________ ___________ ___________ ___________

2. Round the numbers to the nearest hundred.

2851 is _____	6192 is _____	5156 is _____
2528 is _____	5565 is _____	9323 is _____
9134 is _____	9637 is _____	4284 is _____
1781 is _____	3554 is _____	7356 is _____
9276 is _____	7197 is _____	4857 is _____
4612 is _____	4861 is _____	9551 is _____

1. Write the missing numbers and an equation. Subtract or add.

Use Round-Up-Strategy.

77 + 15		48 - 19		18 + 67		57 - 29	
__	__	__	__	__	__	__	__
tens	ones	tens	ones	tens	ones	tens	ones
__		__	__	__		__	__
__	__	__	__	__	__	__	__
+ __	__	- __	__	+ __	__	- __	__
__	__	__	__	__	__	__	__

77 + 15 = ____ 48 – 19 = ___ 18 + 67 = ____ 57 – 29 = ___

__________ __________ __________ __________

2. Add.

250 + 175 + 350 = __

425 + 250 + 375 = __

325 + 225 + 475 = __

625 + 150 + 255 = __

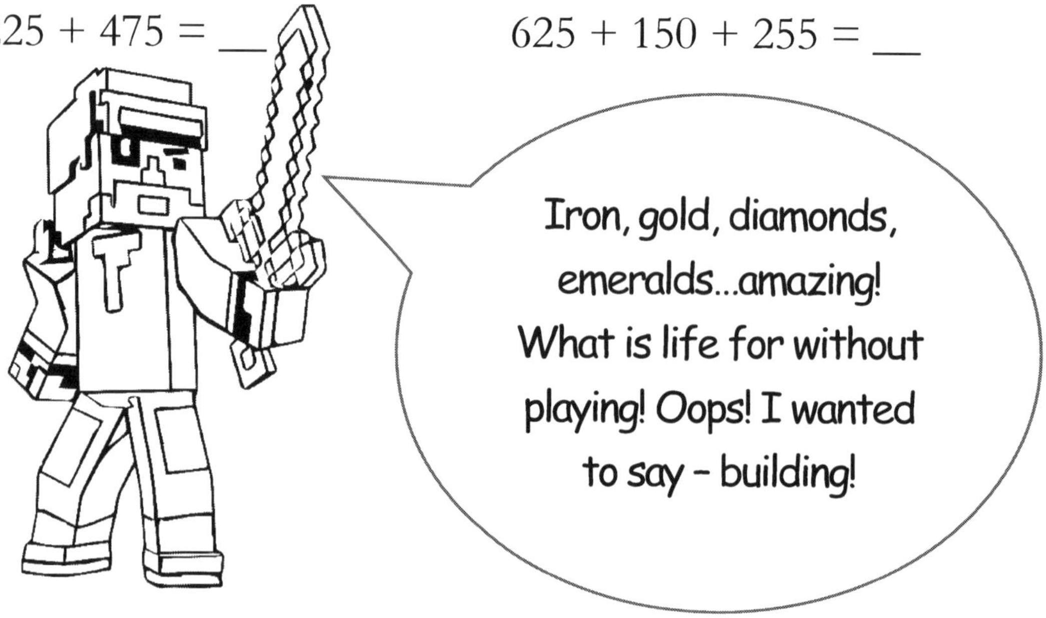

Hint: Ones under ones, tens under tens, hundreds under hundreds;
Subtract ones, then tens, then hundreds.

400 - 76

70 **6**

Decompose 76.
76 is 70+6.

hundreds	**tens**	ones
4	**0**	**0**
-	**7**	**6**
__	__	__

I cannot subtract 6 ones from 0 ones: 0 – 6 = ?!?
Or 7 tens from 0 tens. You must get more ones and more tens.

400-76=400-70-6=330-6=324 or
400-76=400-80+4=320+4=324. I solved it! High five!

Step 1: I subtract in columns, I need one more row above 400.

Step 2: Borrow 1 ten of the 40 tens. Write 10 ones above 0 ones since 1 ten equals 10 ones. Write 39 above 40 tens: 40-1=39.

Step 3: Cross out 0 ones and 40 tens.

hundreds	tens	ones
3	**9**	**10**
4	**~~0~~**	**~~0~~**
-	**7**	**6**
__	__	__

Step 4: Subtract 6 ones from 10 ones: 10-6=4.

Step 5: Subtract 7 tens from 9 tens:9-7=2.

Step 6: Rewrite 3 since 3 – 0 = 3.

hundreds	tens	ones
3	**9**	**10**
4	**~~0~~**	**~~0~~**
-	**7**	**6**
3	**2**	**4**

1. Subtract. Fill in the missing numbers.

96-41=96-40-__	93-__=93-__+__	72-__=72-__-__
35 / 6 / ?...	11 / 27 / ?_	19 / 25 / ?_

2. Subtract.

600 - 71 → 70, __

Hs	Ts	Os
5	9	10
~~6~~	~~0~~	~~0~~
-	7	1
5	2	9

500 - 48 → 40, __

Hs	Ts	Os
__	__	__
__	__	__
-	__	__
__	__	__

300 - 66 → __, __

Hs	Ts	Os
__	__	__
__	__	__
-	__	__
__	__	__

600 - 71 = ______ 500 – 48 = ______ 300 - 66 = ______

______________ ______________ ______________

1. <u>Fill in</u> the missing whole numbers.

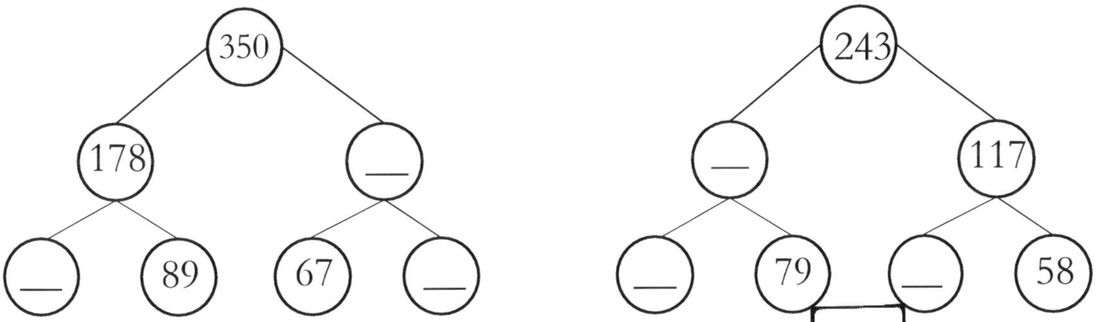

2. <u>Unscramble</u> each word to find out more about the Nether. <u>Write</u> the word in the squares.

tresofrs

gmmaa

idrgeb

inopot

toparl

1. Add.

492 + 531	847 + 225	426 + 158	722 + 536
686 + 153	564 + 333	252 + 175	296 + 188

2. Round the numbers to the nearest hundred.

3497 is ______	1025 is ______	5402 is ______
8214 is ______	7943 is ______	7605 is ______
9734 is ______	3059 is ______	9124 is ______
8246 is ______	4167 is ______	5426 is ______
9214 is ______	9802 is ______	1105 is ______
7635 is ______	7319 is ______	7631 is ______

1. Subtract. Follow the strategy.

500 - 82	700 - 37	400 - 65
90 ___	40 ___	___ ___
Hs Ts Os	Hs Ts Os	Hs Ts Os
___ ___ ___	___ ___ ___	___ ___ ___
___ ___ ___	___ ___ ___	___ ___ ___
- ___ ___	- ___ ___	- ___ ___
___ ___ ___	___ ___ ___	___ ___ ___

500 - 82 = ______

700 – 37 = ______

400 - 65 = ______

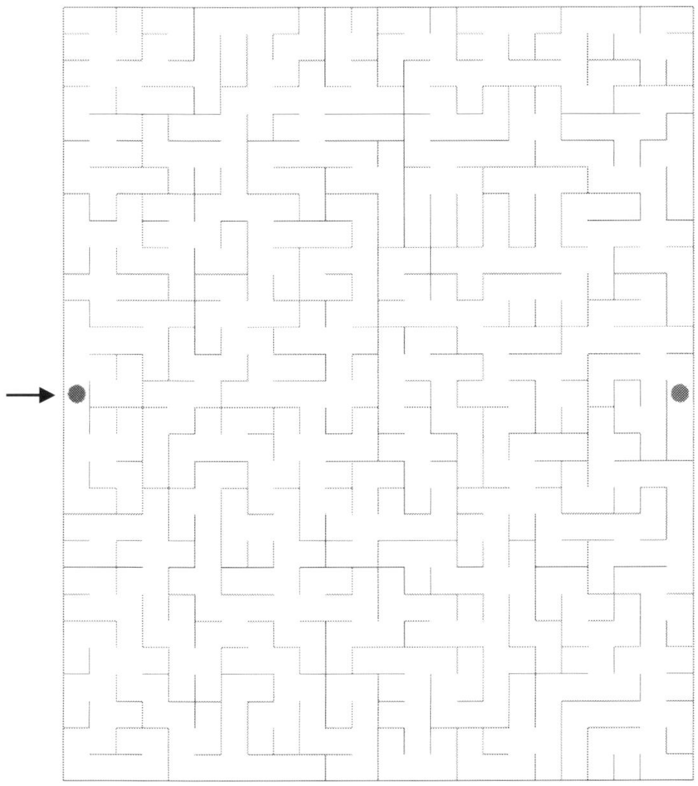

CHAPTER 2

Fun with Renaming/Regrouping

Adding 3-Digit Numbers Vertically up to 1000

Adding 3-Digit Numbers Horizontally up to 1000

Subtracting 3-Digit Numbers Vertically

Subtracting 3-Digit Numbers Horizontally

Adding 3-Digit Numbers Vertically beyond 1000

Adding 3-Digit Numbers Horizontally beyond 1000

Subtracting 4-Digit Numbers Horizontally

Number Talks

Olympiad Math Problems

CogAT test prep

Puzzles

Mazes

Word Search

Cryptograms

1. Write the missing numbers and an equation. Hint: use Round-Up Strategy.

700 - 37

40 3

Hs	Ts	Os
__	__	__
__	__	__
-	__	__
__	__	__

700 - 37 = ______

____

900 - 58

60 __

Hs	Ts	Os
__	__	__
__	__	__
-	__	__
__	__	__

900 – 58 = ______

____

400 - 83

90 __

Hs	Ts	Os
__	__	__
__	__	__
-	__	__
__	__	__

400 - 83 = ______

____

2. Subtract or add.

362 – 8 = __

874 – 9 = __

965 – 7 = __

84 + 37 = __

95 + 68 = __

76 + 69 = __

700 + 240 = __

600 + 330 = __

300 + 480 = __

400 – 225 = __

300 – 175 = __

800 – 455 = __

500 – 285 = __

900 – 745 = __

600 – 395 = __

1. Write the missing numbers and an equation. Hint: use Round-Up Strategy.

400 - 35			600 - 94			900 - 51		
___	___		___	___		___	___	
Hs	Ts	Os	Hs	Ts	Os	Hs	Ts	Os
___	___	___	___	___	___	___	___	___
___	___	___	___	___	___	___	___	___
-	___	___	-	___	___	-	___	___
___	___	___	___	___	___	___	___	___

400 - 35 = ______ 600 – 94 = ______ 900 - 51 = ______

________________ ________________ ________________

2. Fill in the missing numbers.

```
   6  __
-  __  7
---------
   2  6
```

Hm…think what number minus 7 gives us 6?

13, write 3 in the top row. Remember 1 because you subtract 7 out of 3. Subtract 6 – 1 - 2 to get the number in the second row. Got it?

3. Continue a series of numbers:

61, 74, 87, _____, _____, _____, _____, _____, _____

1. Do you like a challenge? I DO! Let's take a timer and see who will solve these problems faster. Adults are welcome, too!

Name ______________________________ Time ____ : ____.

9 + _ = 11	8 + _ = 11	7 + _ = 11	6 + _ = 11
9 + _ = 12	8 + _ = 12	7 + _ = 12	6 + _ = 12
9 + _ = 13	8 + _ = 13	7 + _ = 13	6 + _ = 13
9 + _ = 14	8 + _ = 14	7 + _ = 14	6 + _ = 14
9 + _ = 15	8 + _ = 15	7 + _ = 15	6 + _ = 15
9 + _ = 16	8 + _ = 16	7 + _ = 16	9 + _ = 17
8 + _ = 17	9 + _ = 18	18– _ = 9	17 – _ = 9
17 – _ = 8	16 – _ = 9	16 – _ = 8	16 – _ = 7
15 – _ = 9	15 – _ = 8	15 – _ = 7	15 – _ = 6
14 – _ = 9	14 – _ = 8	14 – _ = 7	14 – _ = 6
14 – _ = 5	13 – _ = 9	13 – _ = 8	13 – _ = 7
13 – _ = 6	13 – _ = 5	13 – _ = 4	12 – _ = 9
12 – _ = 8	12 – _ = 7	12 – _ = 6	12 – _ = 5
12 – _ = 4	12 – _ = 3	11 – _ = 9	11 – _ = 8
11 – _ = 7	11 – _ = 6	11 – _ = 5	11 – _ = 4

2. Continue a series of numbers:

909, 808, _____, _____, _____ , _____, _____, _____

1. Find and circle or cross out the words to find out more about the Ender Dragon.

C	R	Y	S	T	A	L	S	S	H
U	W	D	P	M	P	I	L	G	E
I	Z	P	E	U	E	L	N	G	A
I	K	I	R	S	A	L	V	E	L
M	F	P	G	B	T	G	O	V	I
L	L	C	W	D	V	R	P	G	N
E	G	O	C	Y	M	B	O	P	G
D	N	Y	K	C	I	R	T	Y	H
S	S	O	T	Z	F	N	T	U	F
J	R	A	S	Y	W	N	C	O	L

CRYSTALS
PURPLE
DESTROY
SNOWBALLS
EGGS
TRICKY
GOLEMS
HEALING

Hint: Write ones under ones, tens under tens, and so on; Add ones first, then tens, then hundreds; Carry when the sum in any column is 10 or more.

526 + 175

4 171

Decompose 175. 175 is 4+171.

	hundreds	tens	ones
	5	2	6
+	1	7	5
	__	__	__

6+5=11. So, you write 1 in one's place and carry 1 ten with the other tens.

Aha, I need to add one more row to write the numbers that were carried, right?

	hundreds	tens	ones
	__	__	
	5	2	6
+	1	7	5
	__	__	__

Step 1: I need one more row above 526.

Step 2: First, I add ones: 6+5=11.

Write the 1 in one's place. Carry 1 ten with the other tens. Write the 1 above 2 in ten's place.

	hundreds	tens	ones
	__	1	
	5	2	6
+	1	7	5
	__	__	1

Step 3: Add tens: 1+2+7=10. Write the 0 in ten's place. Carry 1 hundred with the other hundreds. Write the 1 above 5 in hundred's place.

Step 4. Add hundreds: 1+5+1=7.

	hundreds	tens	ones
	1	1	
	5	2	5
+	1	7	5
	7	0	1

1. Write the missing numbers and an equation. Decompose.

397 + 65	481 - 79	382 + 58
6 __	71 __	18 __

Hs	Ts	Os		Hs	Ts	Os		Hs	Ts	Os
__				__	__	__		__		
__	__	__		__	__	__		__	__	__
+	__	__		-	__	__		+	__	__
__	__	__		__	__	__		__	__	__

397 + 65 = ______　　481 – 79 = ______　　382 + 58 = ____

______________　　______________　　______________

2. Round the numbers to the nearest hundred.

451	500	516	__	623	__	782	__	843	__	994	__
427	__	569	__	657	__	796	__	865	__	938	__
494	__	584	__	615	__	744	__	811	__	962	__
448	__	507	__	608	__	729	__	888	__	927	__

3. Continue a series of numbers:

4, 6, 10, 16, ______, ______, ______, ______, ______, ______

1. Evaluate each expression.

28 + 7 – 15 = __ 37 – 9 + 22 = __ 66 + 5 – 23 = __

82 – 6 + 14 = __ 54 + 8 – 17 = __ 74 – 7 + 11 = __

42 + 8 – 36 = __ 70 – 6 + 16 = __ 62 + 9 – 43 = __

2. Add.

Hint: 250+650 = 650+50+200 or 200+600+50+50 or 250+50+600.

150 + 450 = __ 350 + 350 = __ 450 + 550 = __

150 + 850 = __ 750 + 50 = __ 350 + 650 = __

550 + 150 = __ 850 + 50 = __ 350 + 450 = __

450 + 450 = __ 250 + 150 = __ 150 + 750 = __

250 + 550 = __ 450 + 250 = __ 650 + 50 = __

3. Find the value.

99

45 | ? __ | 16

? ___

12 | 27 | 39

? ___

44 | 19 | 25

1. Subtract.

500	800	507	592
- 225	- 635	- 369	- 386
____	____	____	____

300	400	1000	1000
- 175	- 385	- 565	- 397
____	____	____	____

2. Write the missing numbers to make the equations true.

Y___ + Y ___ + Y ___ = 36

Y ___ + B ___ + Y ___ = 33

Y ___ - B ___ + G ___ = 22

3. Fill in the missing numbers.

4 ___	___ 2	___ 3
+ ___ 9	+ 2 8	- 2 ___
9 6	8 ___	5 7

1. Subtract. Hint: use Round-Up-Strategy.

500 – 54 = ___ 200 – 68 = ___ 900 – 83 = ___

400 – 16 = ___ 700 – 39 = ___ 600 – 42 = ___

2. Add. Hint: decompose.

14 + 65 = ___ 26 + 34 = ___ 28 + 59 = ___

57 + 25 = ___ 27 + 64 = ___ 17 + 35 = ___

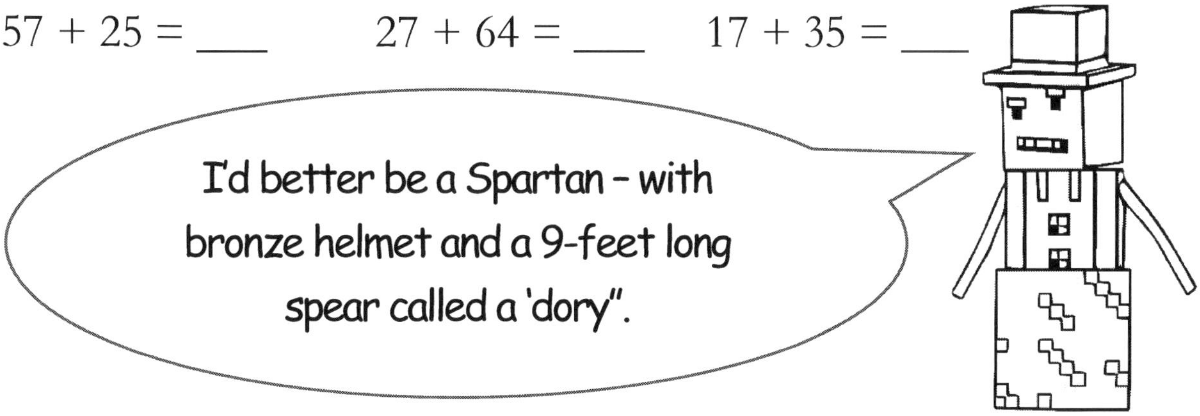

3. Subtract.

245 - 176

145 31

	Hs	Ts	Os
	1	13	15
	~~2~~	4	~~5~~
-	1	7	6
		…	…

245 – 176 = 245 – 145 – 31 = 69

364 - 219

__ __

	Hs	Ts	Os
		__	__
	__	__	__
-	__	__	__
	__	__	__

364 – 219 = ___ – ___ – ___ = ___

853 - 466

__ __

	Hs	Ts	Os
	__	__	__
	__	__	__
-	__	__	__
	__	__	__

853 – 466 = ___ – ___ – __ = ___

1. Find the value. Use the alphabet on page "2" of the workbook to find out the hidden word.

100 - 47 - 35 + 28 – 20 = ______________________________

100 - 53 - 16 + 36 - 58 = ______________________________

100 - 39 - 13 – 23 - 18 = ______________________________

100 - 29 - 38 + 13 - 20 = ______________________________

100 - 45 - 24 + 11 – 41 = ______________________________

100 - 19 - 28 - 36 – 10 = ______________________________

____	____	____	____	____	____
____	____	____	____	____	____

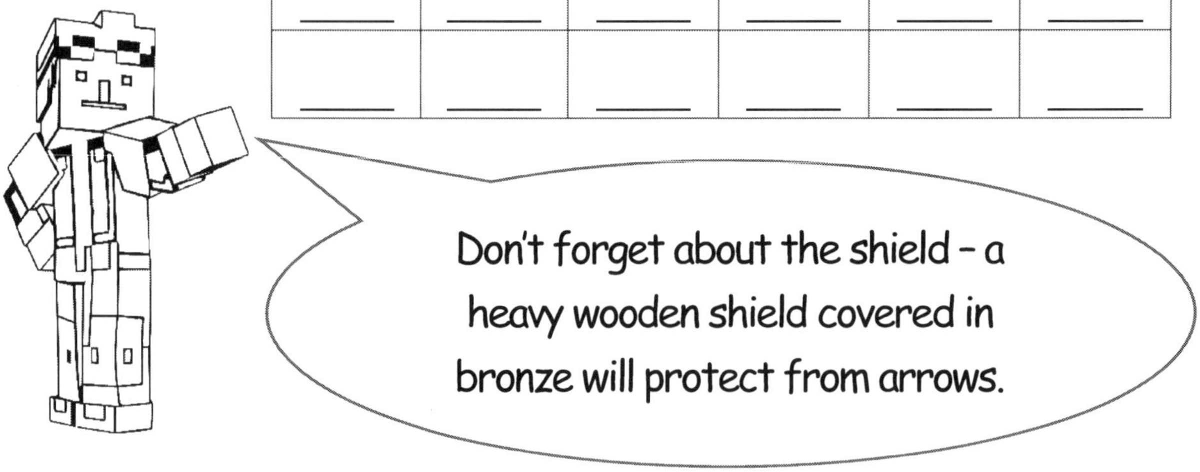

2. Subtract.

300 – 78 = ___   600 – 36 = ___   800 – 59 = ___

3. Continue a series of numbers:

4, 4, 8, 12, _____, _____, _____, _____, _____, _____

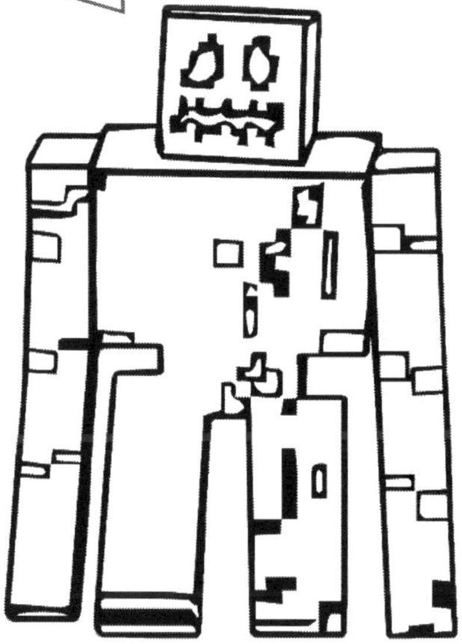

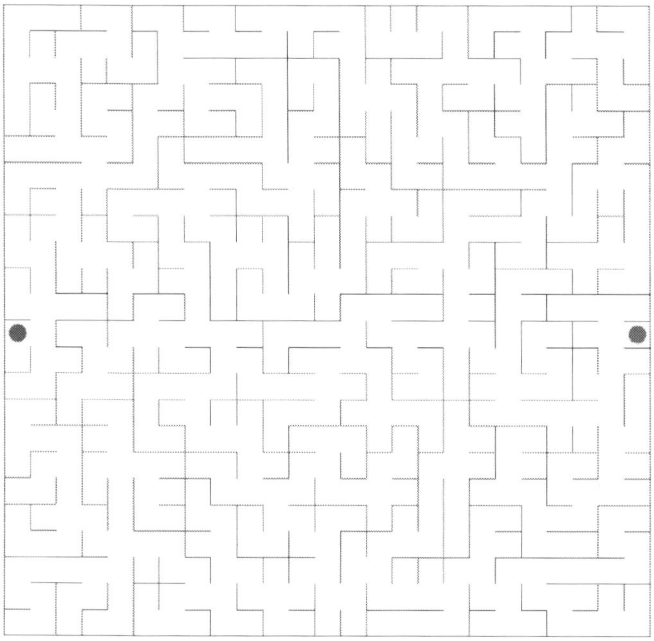

1. Unscramble each word to find out more about the fireworks in Minecraft. Write the word in the squares.

eexdlop

gbna

ewinklt

etocrk

uwroegdnp

1. Fill in the numbers 20, 22, 24, 26, 28, 30, 32, 34, 36 into the squares so that the sum in each row and in each column is 84.

Draw the arrows from the least number to a greater number up to the greatest number: 20 - 22 - 24, etc.

____	____	____
____	____	____
____	____	____

2. Continue a series of numbers:

225, 275, 250, ____, ____, ____, ____, ____, ____, ____, ___

3. Add.

6	7	5	8
10	80	90	50
+ 267	+ 184	+ 232	+ 294
____	____	____	____

4	3	9	7
26	87	74	74
+ 615	+ 429	+ 526	+ 385
____	____	____	____

1. Solve the pattern for each rectangle and find the missing number.

16		15	14		37
	53			20	
13		9	19		12

17		35	15		27
	11			___	
23		18	12		29

2. Fill in the missing numbers and answer the question.

I have __ __ bricks. How many more bricks could I add to change the tens by 1 digit up? Find as many options as you can.

____

____

______________________.

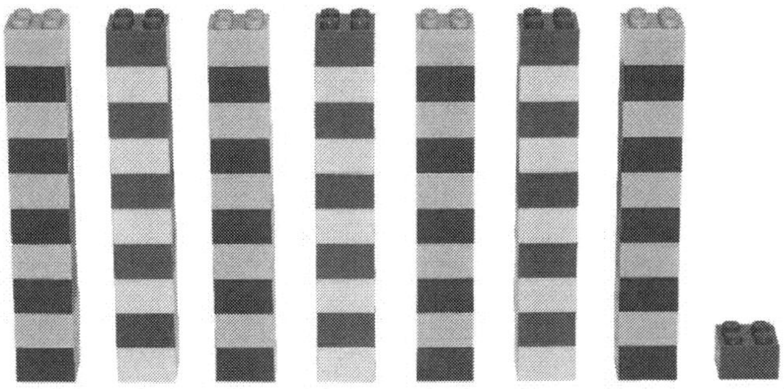

3. Continue a series of numbers:

15, 30, ______, ______, ______, ______, ______, ______.

1. Add.

496 + 158	459 + 377	584 + 149
4 154	__ __	__ __
Hs Ts Os	Hs Ts Os	Hs Ts Os
__ __	__ __	__ __
__ __ __	__ __ __	__ __ __
+ __ __ __	+ __ __ __	+ __ __ __
__ __ __	__ __ __	__ __ __
496 + 158 = _____	459 + 377 = _____	584 + 149= _____
______________	______________	______________

2. Round the numbers to the nearest hundred.

256	300	335	__	401	__	545	__	623	__	779	__
241	__	372	__	464	__	599	__	658	__	794	__
293	__	327	__	438	__	553	__	632	__	735	__
218	__	389	__	470	__	576	__	610	__	797	__

3. Continue a series of numbers:

63, 61, 58, 54, _____, _____, _____, _____, _____, _____.

1. Write "+" or "-" to make the equations true.

56 19 27 = 48 | 34 17 25 = 42

100 64 18 = 54 | 69 38 24 = 7

2. Write the missing numbers and answer the question.

I have ___ ___ bricks. How many bricks could I subtract to change the tens by 1 digit down? Find as many options as you can.

____

____

_______________________.

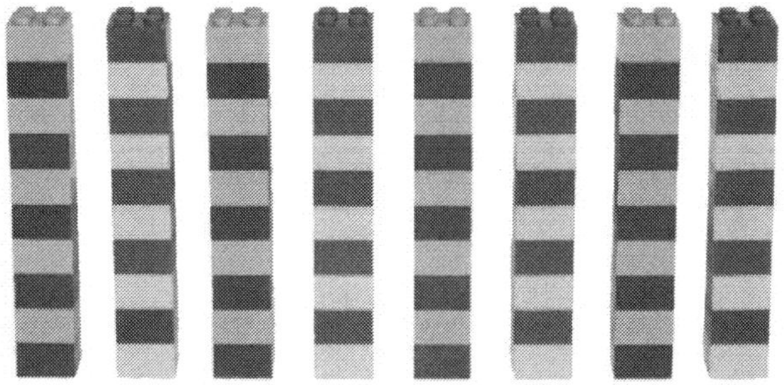

3. Add and compare the sums. Hint: 465 + 200 - 11.

465 + 189 = ___ ___ 314 + 295 = ___

257 + 497 = ___ ___ 164 + 298 = ___

4. Subtract.

1000 – 620 = ___ 1000 – 490 = ___ 1000 – 170 = ___

1. Write the missing numbers.

45 + ? __ = 75	? __ + 5 = 60	85 – ? __ = 38
24 + ? __ = 65	? __ + 30 = 98	90 – ? __ = 51
28 + ? __ = 67	? __ + 22 = 49	65 – ? __ = 18
24 + ? __ = 45	? __ + 19 = 50	71 – ? __ = 34
37 + ? __ = 66	? __ + 12 = 54	57 – ? __ = 29
48 + ? __ = 89	? __ + 28 = 71	44 – ? __ = 18

2. Write 3 different strategies to add and subtract: Decompose, Round-Up, Tens+Ones.

44+25+17= ______________=______________=______________

65+14+29= ______________=______________=______________

44+48–17= ______________=______________=______________

65+17-29= ______________=______________=______________

1. Add or subtract

563 + 128	534 - 228	391 + 199
↙ ↘	↙ ↘	↙ ↘
130 2	___ ___	___ ___

	Hs	Ts	Os
	___	___	
	___	___	___
+	___	___	___
	___	___	___

	Hs	Ts	Os
	___	___	___
	___	___	___
-	___	___	___
	___	___	___

	Hs	Ts	Os
	___	___	
	___	___	___
+	___	___	___
	___	___	___

563 + 128 = ______ 534 – 228 = _____ 391 + 199 = ______

__________________ __________________ __________________

2. Add. Write 2 strategies for each problem. The first one is done for you.

35 + 15 + 26 = 30 + 10 + 20 + 10 + 6 = 40 + 10 + 26 = 76

36 + 44 + 17 = ____________________ = ____________________

55 + 25 + 19 = ____________________ = ____________________

19 + 44 + 28 = ____________________ = ____________________

47 + 14 + 28 = ____________________ = ____________________

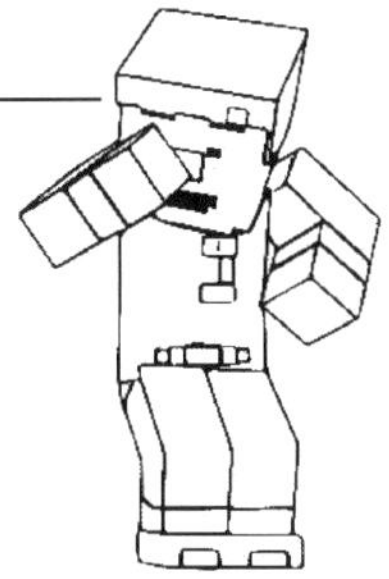

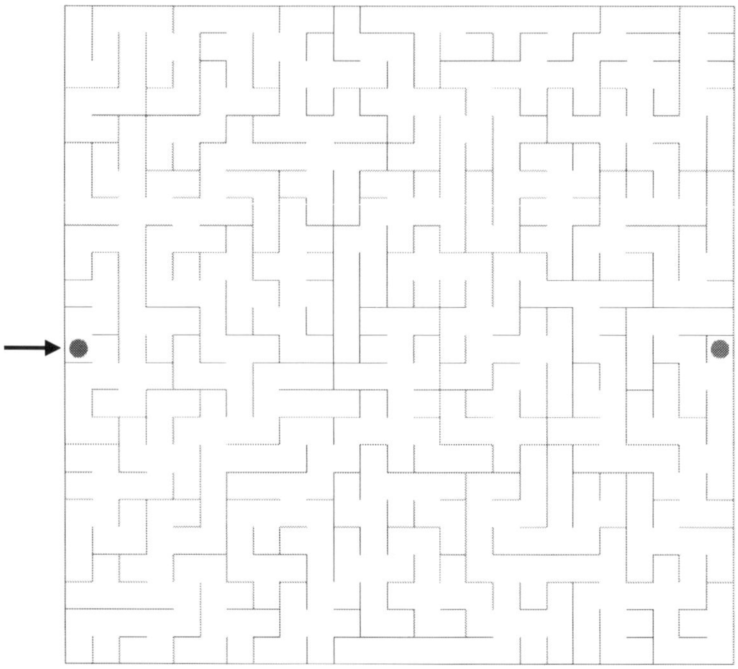

1. Find and circle or cross out the words to find out more about maps in Minecraft.

O	S	R	K	W	K	E	B	N	S
R	V	T	S	O	C	E	O	C	U
H	J	Z	A	W	M	I	O	C	R
A	D	V	E	N	T	U	R	E	V
N	M	K	U	A	D	E	C	P	I
F	M	O	V	Y	A	A	U	P	V
M	P	E	O	T	X	J	R	F	A
Q	L	K	I	Z	A	A	M	D	L
E	O	V	E	M	A	R	F	Y	J
E	E	J	D	D	T	B	A	B	R

ELEVATION
STANDARD
ZOOM
FRAME
ADVENTURE
SURVIVAL
CREATIVE

1. Add. Follow the strategy.

239 + 263	674 + 177	448 + 396
300 __	200 __	__ __
Hs Ts Os	Hs Ts Os	Hs Ts Os
__ __	__ __	__ __
__ __ __	__ __ __	__ __ __
+ __ __ __	+ __ __ __	+ __ __ __
__ __ __	__ __ __	__ __ __

239 + 263 = _____   674 + 177 = _____   448 + 396 = _____

_______________   _______________   _______________

2. Add. Draw the arrows to show easier calculations.

91 + 55 + 22 + 15 + 19 + 48 =

____

56 + 26 + 23 + 14 + 17 + 44 =

____

24 + 27 + 21 + 23 + 26 + 29 =

____

3. Subtract.

1000 – 540 = ___   1000 – 360 = ___   1000 – 790 = ___

1. <u>Evaluate</u> each equation. <u>Use</u> the alphabet on page "2" of the workbook to find out the hidden word.

28 + 33 + 29 – 64 – 20 = ______________________________

23 + 59 + 2 – 37 - 16 = ______________________________

14 + 18 + 29 – 28 – 21 = ______________________________

36 + 17 + 16 – 24 – 26 = ______________________________

52 + 17 + 19 – 22 – 41 = ______________________________

What is MY favorite word?☹

__	__	__	__	__
__	__	__	__	__

2. <u>Write</u> the missing numbers to make the sum and difference true.

```
     6  __          3  __  6
+  __   3       -      3  __
----------      -------------
__  1   0          __  1  8
```

Boy, I'd dream to build a wooden Trojan Horse! Just imagine - when I see skeletons, I build a Trojan Horse.

1. Evaluate each equation. Use the alphabet on page "2" of the workbook to find out the hidden word.

71 + 24 - 18 – 15 – 58 = ______________________________

12 + 29 + 17 – 26 - 23 = ______________________________

53 + 16 + 17 – 43 – 17 = ______________________________

41 + 15 + 18 – 36 – 12 = ______________________________

80 - 14 - 19 + 8 – 30 = ______________________________

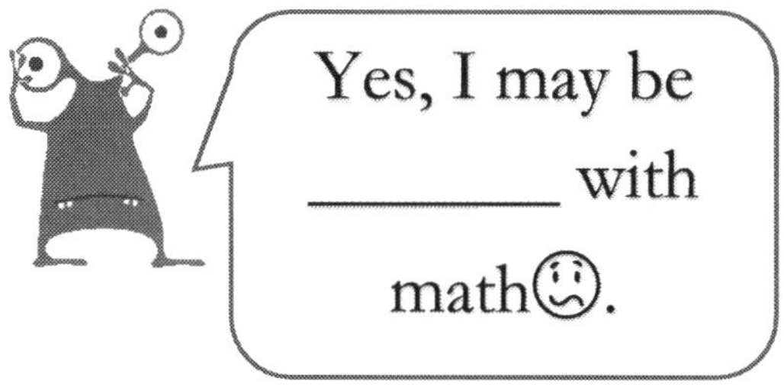

__	__	__	__	__
__	__	__	__	__

2. Continue the pattern:

1999, 1888, 1666, _______, _______, _______

3. Add and subtract.

800 – 458 = ___

135 + 835 = ___

700 – 282 = ___

1000 – 225 = ___

1000 – 328 = ___

1000 – 768 = ___

1. Subtract.

381 - 193 → 181 ___

Hs Ts Os

	Hs	Ts	Os
	__	__	__
	__	__	__
-	__	__	__
	__	__	__

381 – 193 = ___ – ___ – ___ = ___

435 - 157 → ___ ___

	Hs	Ts	Os
	__	__	__
	__	__	__
-	__	__	__
	__	__	__

435 – 157 = _____ ________________

574 - 186 → ___ ___

	Hs	Ts	Os
	__	__	__
	__	__	__
-	__	__	__
	__	__	__

574 – 186 = _____ ________________

2. Add.

Hint: 154 + 67 = 154 + 100 - 33 or 154 + 6 + 60 + 1☺.

142 + 89 = ________________________________

126 + 75 = ________________________________

135 + 39 = ________________________________

3. Write the missing numbers to make the sum and difference true.

	8	0	__
-	4	__	5
	__	4	5

	1	__	__
+	4	8	8
	__	5	1

1. Add.

189 + 146	198 + 374	265 + 247
11 135	__ __	__ __
Hs Ts Os	Hs Ts Os	Hs Ts Os
__ __	__ __	__ __
__ __ __	__ __ __	__ __ __
+ __ __ __	+ __ __ __	+ __ __ __
__ __ __	__ __ __	__ __ __
189 + 146 = ______	374 + 198 = ______	265 + 247= ____
______________	______________	______________

Why do I decompose 374, but not 198?

____

____.

2. Add.

125 + 275 = ___

375 + 225 = ___

575 + 25 = ___

175 + 425 = ___

725 + 175 = ___

425 + 75 = ___

1. Add.

360	540	429	731
+ 280	+ 160	+ 150	+ 190
______	______	______	______

264	582	198	356
+ 139	+ 159	+ 278	+ 268
______	______	______	______

2. Add. Draw the arrows to show the best strategy.

44 + 23 + 18 + 27 + 16 + 42 = ________________________

55 + 12 + 21 + 36 + 15 + 19 + 18 + 24 = ________________

3. Add or subtract. Hint: use Round-Up Strategy.

424 + 199 = ____________

314 + 295 = ____________

857 – 397 = ____________

564 – 298 = ____________

684 – 195 = ____________

578 + 391 = ____________

1. Unscramble each word to find out more about the ocean. Write the word in the squares.

istelho

sidanls

wsahlol

acyl

aes elmpte

1. Evaluate each expression.

88 – 39 + 22 – 44 + 58 = ____________________

39 + 32 – 46 + 27 – 38 = ____________________

58 + 19 – 49 + 15 – 27 = ____________________

63 – 37 + 45 – 12 + 24 = ____________________

2. Subtract.

524	437	814	571
70	60	90	80
- 5	- 8	- 7	- 6
____	____	____	____

657	384	425	362
38	59	76	93
- 5	- 8	- 4	- 6
____	____	____	____

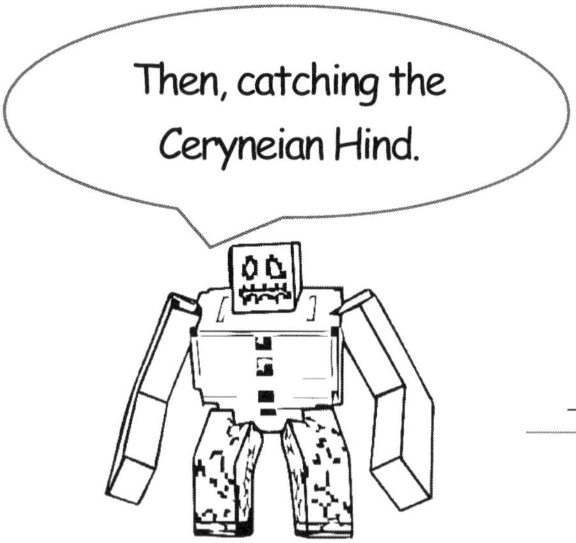

3. Write the missing numbers to make the sum and difference true.

6 0 __		4 __ __
- 3 __ 3		+ 1 6 9
__ 3 7		__ 3 0

1. Subtract.

520	640	800	700
- 225	- 375	- 365	- 425
____	____	____	____
813	485	680	612
- 276	- 268	- 483	- 209
____	____	____	____

2. Add. Hint: 75 = 25 + 50.

The smartest Brainer thinks the easiest way is to calculate by 50's or 100's, so, make 50 or 100 with 25 and 75.

525 + 275 = ____   875 + 125 = ____   25 + 775 = ____

425 + 375 = ____   225 + 175 = ____   625 + 175 = ____

175 + 325 = ____   425 + 575 = ____   525 + 175 = ____

3. Subtract. Hint: use Round-Up Strategy.

300 – 146 = _____

600 – 289 = _____

900 – 697 = _____

I remember him finding the Lernaean Hydra!

1. Add. Hint: decompose.

535 + 228 = ___ 805 + 86 = ___ 346 + 357 = ___

486 + 314 = ___ 264 + 226 = ___ 155 + 125 = ___

265 + 217 = ___ 436 + 329 = ___ 697 + 203 = ___

2. Subtract.

	6	1		9	2		9	4		7	3		3	5
	2	2		3	3		3	5		4	4		1	6
-	3	5	-	4	4	-	1	6	-	1	3	-	1	2
	__	__		__	__		__	__		__	__		__	__

	5	5		8	7		5	4		8	1		4	6
	1	7		4	9		2	6		5	3		1	8
-	2	3	-	2	2	-	2	5	-	2	4	-	2	1
	__	__		__	__		__	__		__	__		__	__

	7	7		5	8		6	6		8	2		9	9
		7			8			6			2			9
-	1	7	-	2	6	-	4	2	-	4	8	-	7	1
	__	__		__	__		__	__		__	__		__	__

1. Subtract.

500	800	500	600
- 225	- 635	- 365	- 385
______	______	______	______
300	400	1000	1000
- 175	- 385	- 565	- 395
______	______	______	______

2. Add. Use 2 or 3 different addition Strategies.

386+155 = 155+400–14=386+14+141=380+120+6+35=541

548 + 244 = ____

653 + 197 = ____

365 + 477 = ____

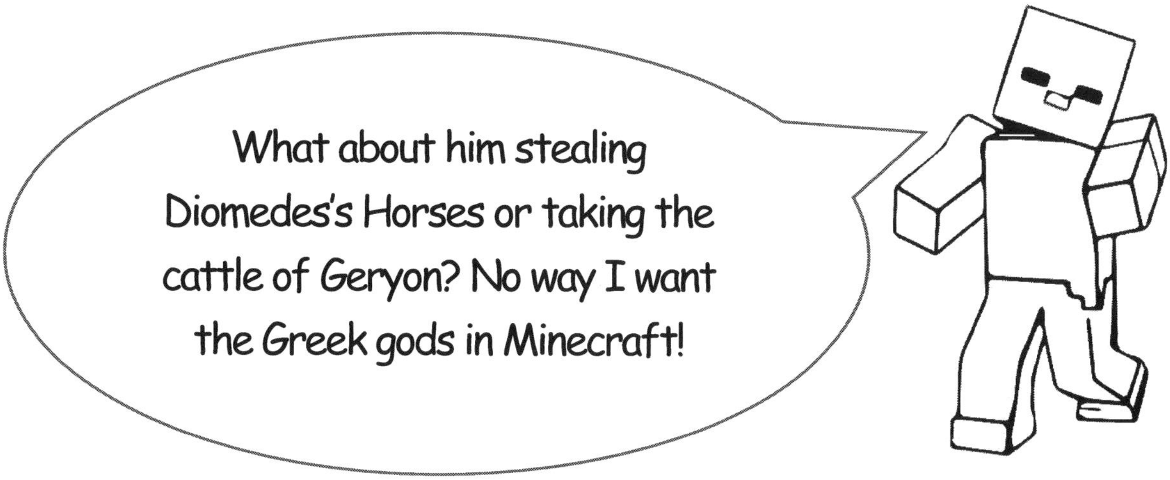

1. Fill in the missing numbers to make the equations true.

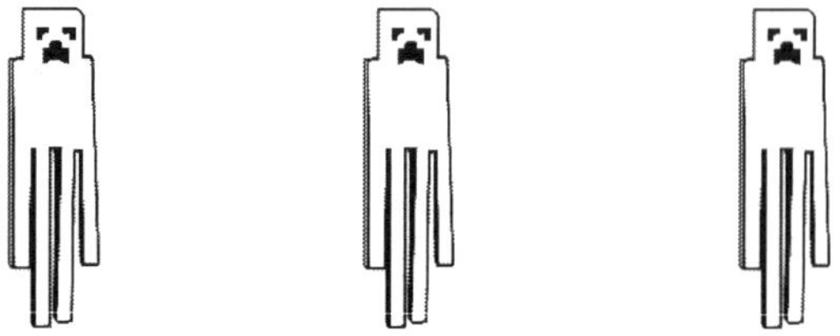

___ + ___ + ___ = 66

___ + ___ + ___ = 72

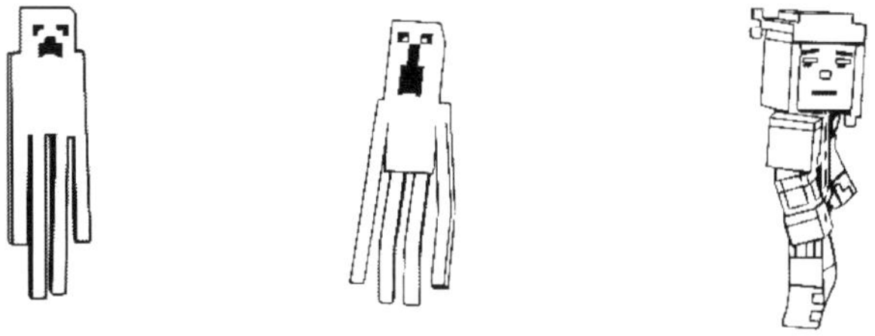

___ + ___ + ___ = 85

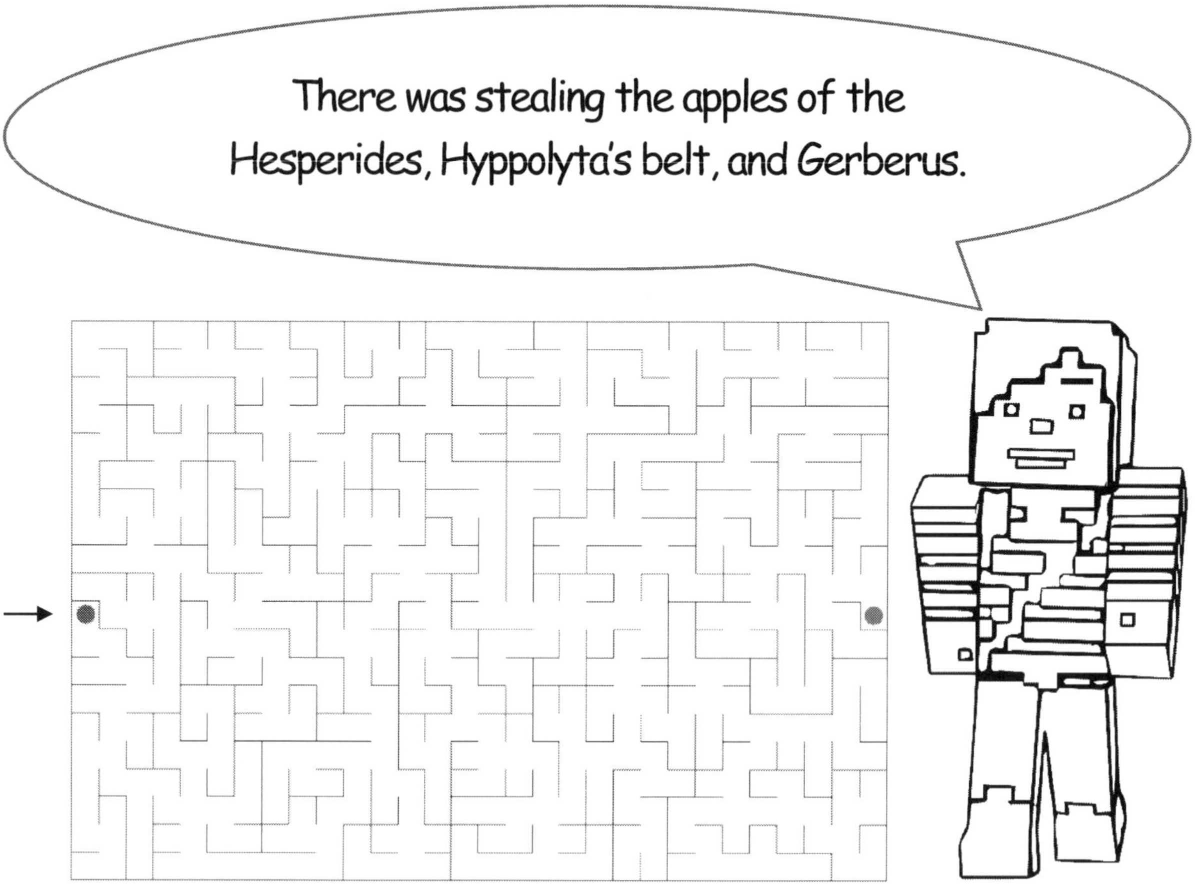

1. Write the missing numbers 275, 300, 650, 850, and 925 to make the equation true.

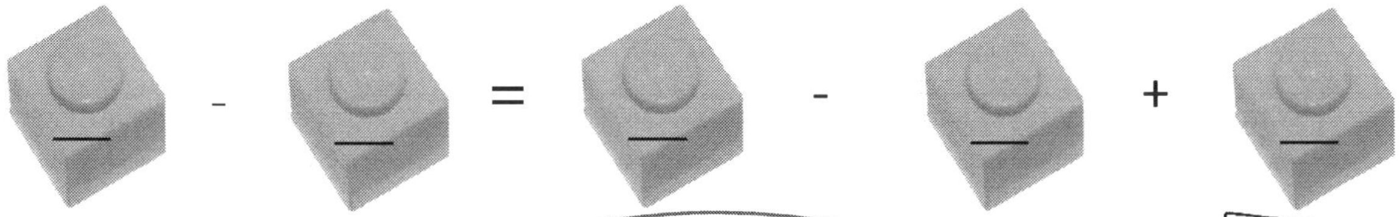

2. Answer the question.

A	B	C	D	E	F	G	H	I	J	K	L	M
1	2	3	4	5	6	7	8	9	10	11	12	13
N	O	P	Q	R	S	T	U	V	W	X	Y	Z
14	15	16	17	18	19	20	21	22	23	24	25	26

__ __ __ __   __ __   __ __ __   __ __ __ __   __ __

23 8 1 20   4 15   25 15 21   14 5 5 4   20 15

__ __ __ __ __   __ __ __ __ __ __ __   __ __ __ __ __?

3 18 1 6 20   19 9 24 20 5 5 14   18 1 9 12 19

____

1. Answer the questions.

If you build bricks in pairs, how many pairs will you make? _____

How many will be left? _____

How many 3s can you build with the bricks? _____

How many will be left? _____

How many 4s can you build with the bricks? _____

How many will be left? _____

How many 5s can you build with the bricks? _____

How many will be left? _____

How many 6s can you build with the bricks? _____

How many will be left? _____

How many 7s can you build with the bricks? _____

How many will be left? _____

How many 8s can you build with the bricks? _____

How many will be left? _____

How many 9s can you build with the bricks? _____

How many will be left? _____

How many 10s can you build with the bricks? _____

How many will be left? _____

CHAPTER 1

Fun with and without Renaming/Regrouping

Adding 2-Digit Numbers Vertically up to 1000

Adding 2-Digit Numbers Horizontally up to 1000

Subtracting 2-Digit Numbers Vertically

Subtracting 2-Digit Numbers Horizontally

Adding 3-Digit Numbers Horizontally up to 1000

Subtracting 3-Digit Numbers Horizontally

Number Talks

Olympiad Math Problems

CogAT test prep

Puzzles

Mazes

Word Search

Cryptograms

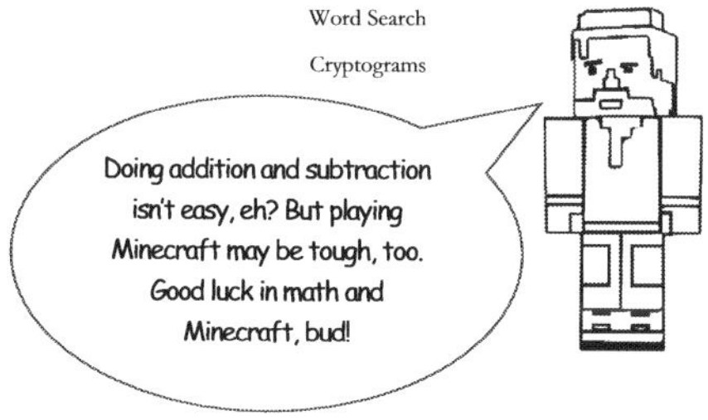

1. Add or subtract.

50 – 6 = 44	60 – 3 = 57	80 – 7 = 73
72 + 8 = 80	87 + 3 = 90	46 + 4 = 50

2. Subtract. Decompose the subtrahend.

91: 66, 25 (70, 4)
73: 25, 48 (30, 5)
83: 48, 35
62: 34, 28
51: 17, 34
100: 36, 64

3. Write the missing numbers and an equation. Subtract.

40 - 27 (30, 3)	27 = 30 - 3; 40 - 30 + 3 = 13	60 - 32 (40, 8)	32 = 40-8; 60-40+8 = 28
80 - 59 (60, 1)	59 = 60-1; 80-60+1=21	50 - 16 (20, 4)	16 = 20-4; 50-20+4 = 34

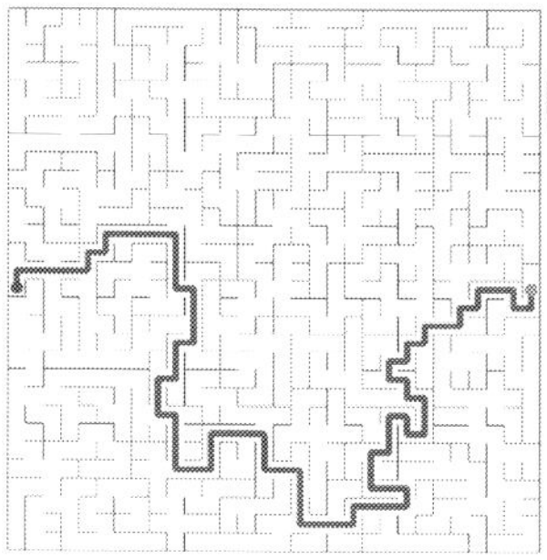

1. Find and circle or cross out the words to find out what you can catch fishing in Minecraft.

S	K	F	S	A	D	D	L	E	S
B	T	F	S	H	H	S	H	P	B
O	P	I	O	K	D	O	S	S	C
W	R	O	C	O	O	S	I	H	B
L	K	T	R	K	E	O	F	B	H
S	U	R	W	L	S	D	B	O	L
O	F	A	T	S	E	N	O	B	H
M	K	T	Y	J	G	P	G	B	R
C	O	P	R	X	K	C	J	P	F
B	D	E	R	S	R	N	Z	D	S

FISH
BOWLS
STICKS
RODS
HOOKS
BOTTLES
BONES
BOOKS
SADDLES

1. Compare and find by how many more or less the numbers are different.

80 > 20 + 20 by	40	100 - 55 < 70 by	25
90 - 60 > 10 by	20	60 < 36 + 44 by	20
20 < 10 + 90 by	80	43 + 27 < 100 by	30
70 - 50 > 10 by	10	30 < 53 + 38 by	61
95 > 30 + 50 by	55	95 - 47 < 64 by	16

2. Subtract. Decompose the subtrahend.

100: 65, 35 (60, 5)
100: 85, 15
100: 38, 62
70: 46, 24
80: 25, 55
60: 41, 19

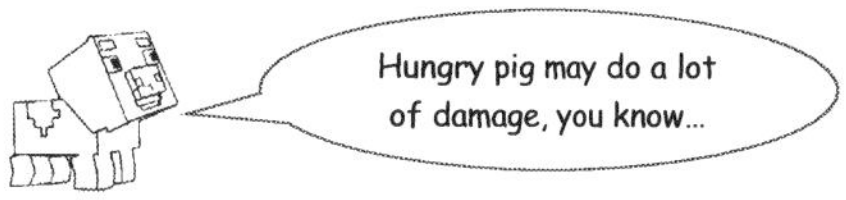

1. Fill in the missing numbers to make the diagrams true.

8 + 1 = 9; 2 + 5 = 7; 9 + 7 = 16

90 - 55 = 35; 24 - 15 = 9; 35 - 9 = 26

2. Subtract.

142 - 89 = 42+11=53

126 - 75 = 26+25=51

131 - 79 = 31+21=52

3. Fill in the missing numbers (__ equals one digit) to make the equations true.

9 9 + 1 = 1 0 0

9 9 9 + 1 = 1 0 0 0

1 0 0 0 0 - 1 = 9 9 9 9

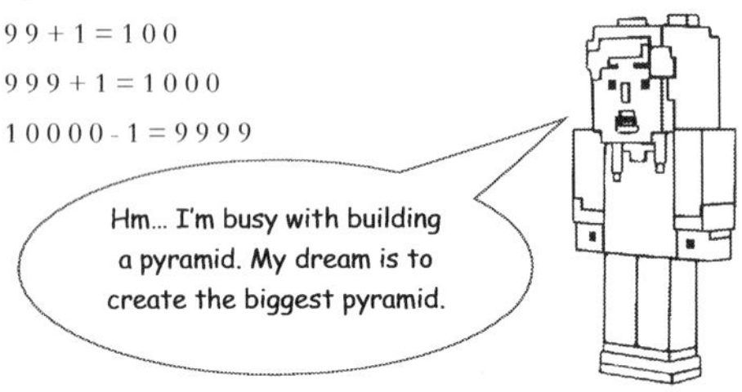

Hint: Borrow from the next figure to the left; When borrowing 1 ten, change it to 10 ones; When borrowing 1 hundred, change it to 10 tens.

72 - 38 → 32, 6

Decompose 38. 38 is 32+6.

tens	ones
7	2
- 3	8
__	__

In columns we subtract 8 ones from 2 ones.

But you cannot subtract 8 ones from 2 ones: 2 – 8 = ?!? You must get more ones.

Wait… 72 - 32 – 6 = 40 - 6 = 34.

Step 1: If I subtract in columns, I need one more row above 72.

tens	ones
__	__
7	2
- 3	8
__	__

Step 2: I borrow 1 ten from 7 tens since 70=10+60.

10 borrowed ones + 2 ones are 12 ones, write 12 above 2 in one's place. Write the leftover 6 tens above 7 in ten's place.

tens	ones
6	12
7	2
- 3	8
__	__

Step 3: Cross out 7 and 2 to avoid mistakes.

Step 4: Subtract 8 ones from 12 ones: 12-8=4.

Step 5: Subtract 3 tens from 6 tens: 6 – 3 = 3.

tens	ones
6	12
~~7~~	~~2~~
- 3	8
3	4

Hint: Write ones under ones, tens under tens, and so on; Add ones first, then tens, then hundreds; Carry when the sum in any column is 10 or more.

44 + 29 → 6, 23

Decompose 29. 29 is 6+23.

tens	ones
4	4
+ 2	9
__	__

4+9=13. So, you write 3 in one's place and carry 1 ten with the other tens.

Aha, I need to add one more row to write the numbers that were carried, right?

Wait… 44 + 29 = 44 + 6 + 23 = 50 + 23 = 73.

tens	ones
__	__
4	4
+ 2	9
__	__

Step 1: If I add in columns, I need one more row above 44.

Step 2: First, I add ones: 4+9=13. Write the 3 in one's place. Carry 1 ten with the other tens. Write the 1 above 4.

tens	ones
1	__
4	4
+ 2	9
__	3

Step 3: Add tens: 1+4+2=7 Write the 7 in ten's place.

tens	ones
1	__
4	4
+ 2	9
5	3

1. Subtract. Use the strategy below: 100–76=100–80+4 = 24.

100-50+5: 45, 55

100-20+5: 15, 85

100-70+6: 64, 36

90-80+3: 77, 13

50-30+5: 25, 25

80-60+5: 55, 25

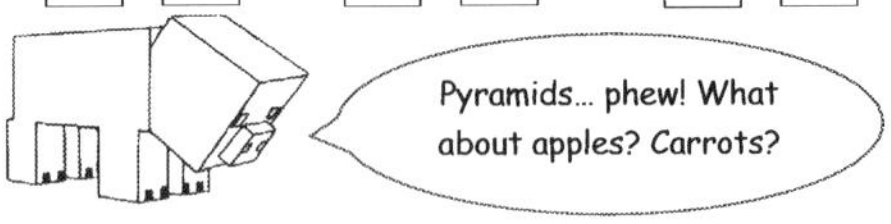

2. Add or subtract.

39 + 13 → 1, 12

tens	ones
1	
3	9
+ 1	3
5	2

43 - 24 → 23, 1

tens	ones
3	13
~~4~~	~~3~~
- 2	4
1	9

29 + 59 → 1, 58

tens	ones
1	
2	9
+ 5	9
8	8

93 - 36 → 33, 3

tens	ones
8	13
~~9~~	~~3~~
- 3	6
5	7

39 + 13 = 39 + 1 + 12 =40+12=52

43 – 24 = 43-23-1=20-1=19

29 + 59 = 29+1+58=88

93 – 36 = 93-33-3=57

1. Circle the right answer.

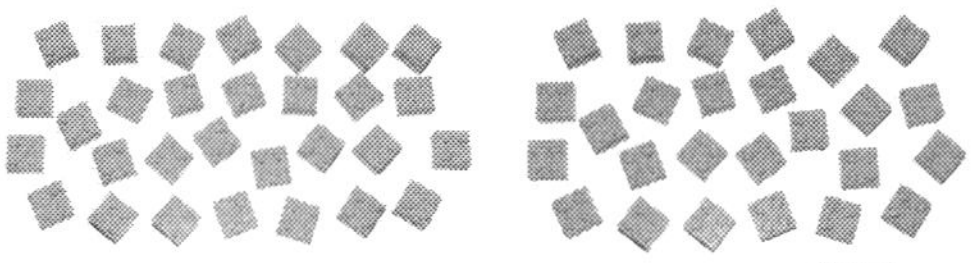
A B

A is greater than B (circled)
A is less than B
A is equal to B

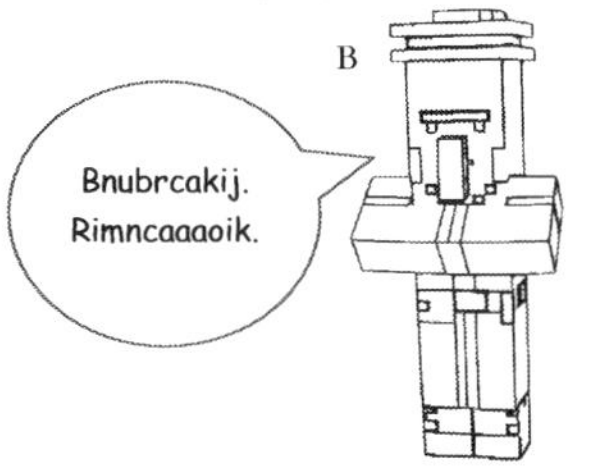

2. Subtract. Follow Round-Up-Strategy.

100-20+5	100-80+4	100-60+5
15 | 85	76 | 24	55 | 45
50-40+2	40-20+2	70-30+1
38 | 12	18 | 22	29 | 41

1. Write the missing numbers and an equation. Add or subtract.
Hint: use Round-Up-Strategy.

46 + 29	51 - 34	68 + 15	94 - 47
30 1	40 6	20 5	50 3
tens ones	tens ones	tens ones	tens ones
1	4 11	1	8 14
4 6	~~5~~ ~~1~~	6 8	~~9~~ ~~4~~
+ 2 9	- 3 4	+ 1 5	- 4 7
7 5	1 7	8 3	4 7

46 + 29 = 46+30-1=75
51 – 34 = 51-40+6=17
68 + 15 = 68+20-5=83
94 – 47 = 94-50+3=47

2. Subtract. Hint: use Round-Up-Strategy.

1000 – 95 = 905 1000 – 82 = 918 1000 – 38 = 962

800 – 64 = 736 800 – 79 = 721 800 – 41 = 759

3. Add or subtract.

? 72	84	67
24 | 37 | 11	26 | 39 | 19	18 | 25 | 24

1. Write the missing numbers 15, 23, 26, 27, and 45.

26 + 15 = 45 - 27 + 23

2. Answer the question.

A	B	C	D	E	F	G	H	I	J	K	L	M
1	2	3	4	5	6	7	8	9	10	11	12	13
N	O	P	Q	R	S	T	U	V	W	X	Y	Z
14	15	16	17	18	19	20	21	22	23	24	25	26

W h a t d o y o u n e e d t o
23 8 1 20 4 15 25 15 21 14 5 5 4 20 15

c r a f t s a n d s t o n e?
3 18 1 6 20 19 1 14 4 19 20 15 14 5

Four blocks of sand.

1. Add or subtract.

58 + 25	83 - 75	38 + 34	82 - 68
2 23	73 2	2 32	62 6
tens ones	tens ones	tens ones	tens ones
1	7 13	1	7 12
5 8	8 3	3 8	8 2
+ 2 5	- 7 5	+ 3 4	- 6 8
8 3	8	7 2	1 4

58 + 25 = 58+2+23=83
83 – 75 = 83-73-2=8
38 + 34 = 38+2+32=72
82 – 68 = 82-62-6=14

2. Fill in the missing numbers ("__" equals one digit).

102 – 9 = 9 3 105 - 7 = 9 8 111 – 1 3 = 9 8

3. The sum of 2 numbers is 40 and their difference is 16. Find these numbers: 12 and 28: 40–16 = 24, 12+12 = 24, 12 + 16 = 28.

4. Continue a series of numbers:

5, 6, 8, 9, 11, 12, 14, 15, 17, 18, 20

1. Fill in the missing whole numbers.

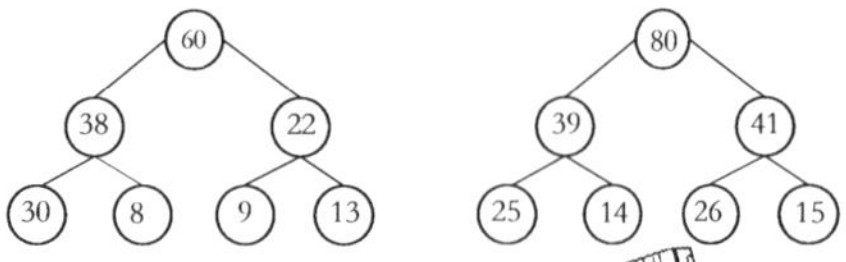

2. Unscramble each word to find out what flowers you can generate in Minecraft. Write the word in the squares.

chrodi	o r c h i d
cors	r o s e
cilla	l i l a c
noildeand	d a n d e l i o n
wrensuflo	s u n f l o w e r

Guys, what is Rounding?!

I LOVE rounding! Rounding is a simpler and shorter way to represent a number. But you need to keep an approximately equal value.

Eh?!.. Do you hear THAT?! Do you have a fever?

Listen, I have $24. I need to round it to the nearest ten. I explore the number range from 20 to 30:

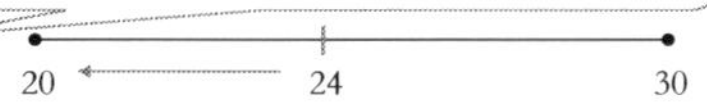

24 is close to 20 since 24 – 20 = 4

24 is not as close to 30 since 30 – 24 = 6

4 < 6, so, 24 is closer to 20 than to 30.

If you round a number followed by 1,2,3, and 4, you round the number down: 24 is 20 or 62 is 60.

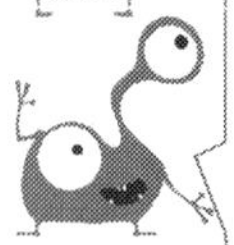

I have $26. I need to round it to the nearest ten.

26 is close to 30 since 30 – 26 = 4

26 is not as close to 20 since 26 – 20 = 6

6 > 4, so, 26 is closer to 30 than to 20.

If you round a number followed by 6,7,8, and 9, you round the number up: 26 is 30 or 87 is 90.

Ha! Looks like these numbers don't like to run☺.

Look, the numbers always run to the distance that is less. The number 24 runs to 20 but not to 30, right?

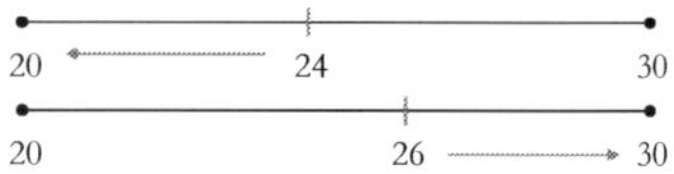

But how can I round $25?

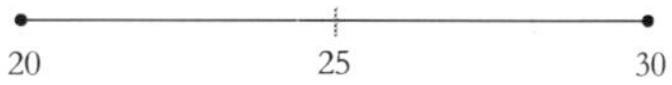

25 is exactly between 20 and 30:

25 is close to 30 since 30 – 25 = 5

25 is also close to 20 since 25 – 20 = 5

25 is Lucky! The numbers ending in 5 always round up! 5 rounds to 10, 15 to 20, 25 to 30, 35 to 40, 45 to 50, 55 to 60, 65 to 70, 75 to 80, 85 to 90, 95 to 100. Done!

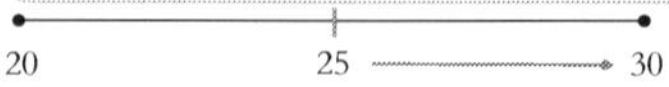

1. Round the numbers to the nearest ten.

89 is 90	63 is 60	47 is 50
22 is 20	34 is 30	75 is 80
91 is 90	56 is 60	18 is 20

2. Add or subtract.

46 + 45	73 - 27	68 + 16	36 - 18
4 41	23 4	20 4	20 2

	tens	ones		tens	ones		tens	ones		tens	ones
	1			6	13		1			2	16
	4	6		~~7~~	~~3~~		6	8		~~3~~	~~6~~
+	4	5	-	2	7	+	1	6	-	1	8
	9	1		4	6		8	4		1	8

46 + 45 = 46+4+41=91

73 – 27 = 73-23-4=46

68 + 16 = 68+20-4=84

36 – 18 = 36-20+2=18

1. Do you like a challenge? I DO! Let's take a timer and see who will solve these problems faster. Adults are welcome, too!

Name ________________________ Time ____ : ____.

9 + 2 = 11	11 – 2 = 9	8 + 3 = 11	11 – 3 = 8
9 + 3 = 12	12 – 3 = 9	8 + 4 = 12	12 – 4 = 8
9 + 4 = 13	13 – 4 = 9	8 + 5 = 13	13 – 5 = 8
9 + 5 = 14	14 – 5 = 9	8 + 6 = 14	14 – 6 = 8
9 + 6 = 15	15 – 6 = 9	8 + 7 = 15	15 – 7 = 8
9 + 7 = 16	16 – 7 = 9	8 + 8 = 16	16 – 8 = 8
9 + 8 = 17	17 – 8 = 9	8 + 9 = 17	17 – 9 = 8
9 + 9 = 18	18 – 9 = 9	6 + 5 = 11	11 – 5 = 6
5 + 6 = 11	11 – 6 = 5	6 + 6 = 12	12 – 6 = 6
5 + 7 = 12	12 – 7 = 5	6 + 7 = 13	13 – 7 = 6
5 + 8 = 13	13 – 8 = 5	6 + 8 = 14	14 – 8 = 6
5 + 9 = 14	14 – 9 = 5	6 + 9 = 15	15 – 9 = 6
7 + 4 = 11	11 – 4 = 7	4 + 7 = 11	11 – 7 = 4
7 + 5 = 12	12 – 5 = 7	4 + 8 = 12	12 – 8 = 4
7 + 6 = 13	13 – 6 = 7	4 + 9 = 13	13 – 9 = 4
7 + 7 = 14	14 – 7 = 7	3 + 8 = 11	11 – 8 = 3
7 + 8 = 15	15 – 8 = 7	3 + 9 = 12	11 – 9 = 2
7 + 9 = 16	16 – 9 = 7	8 + 5 = 13	13 – 5 = 8

1. Round the numbers to the nearest ten.

895 is 900	634 is 630	476 is 480
222 is 220	349 is 350	753 is 750
917 is 920	562 is 560	188 is 190
146 is 150	328 is 330	856 is 860
925 is 930	751 is 750	117 is 120
367 is 370	423 is 420	191 is 190

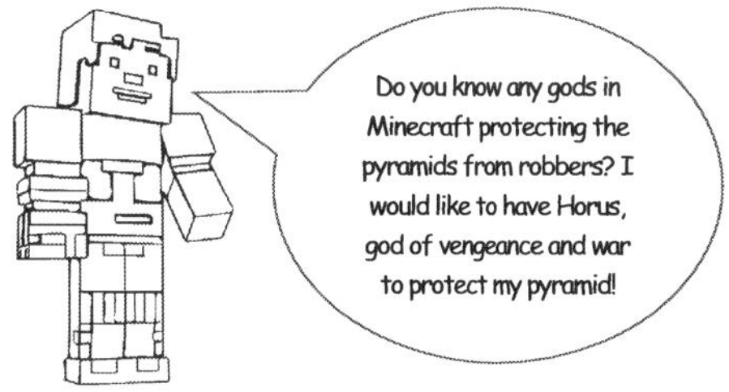

2. Write the missing numbers and an equation. Subtract.

44 - 29
30 1
29 = 30 - 1
44 - 30 + 1 = 15

37 - 19
20 1
19=20-1
37-20+1=18

61 - 35
40 5
35=40-5
61-40+5=26

83 - 46
50 4
46=50-4
83-50+4=37

1. Find the missing number.

100 - X = 67	100 - X = 28	100 - X = 2
X = 33	X = 72	X = 98
100 - ? = 46	100 - ? = 24	100 - ? = 71
? = 54	? = 76	? = 29
100 - ? = 45	100 - ? = 73	100 - ? = 39
? = 55	? = 27	? = 61

2. Round the numbers to the nearest hundred.

140 is 100	370 is 400	590 is 600
220 is 200	860 is 900	630 is 600
910 is 900	480 is 500	750 is 800

3. Find the value. Draw the arrows to make the tens.

3 + 1 + 7 + 6 + 4 + 9 + 4 + 7 + 8 + 12 + 16 + 13 = 90

13 + 11 + 8 + 5 + 4 + 9 + 15 + 7 + 12 + 16 = 100

12 + 15 + 11 + 14 + 17 + 19 + 18 + 13 + 15 + 16 = 150

1. Fill in the missing numbers. A number has to be more than 1.

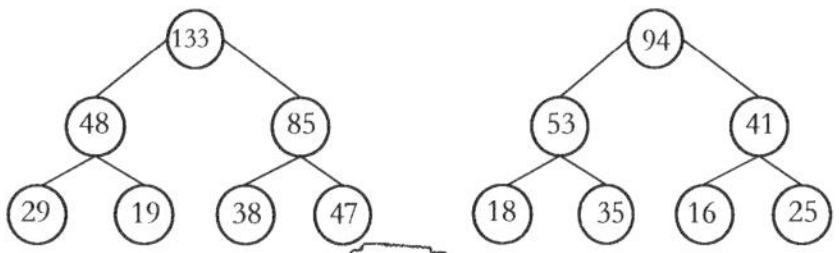

2. Find and circle or cross out the words to find out more about horses in Minecraft.

M	F	M	W	S	A	Q	N	R	B
S	C	T	O	Q	S	I	Z	E	R
G	Q	V	N	Y	K	R	Y	B	E
R	A	G	U	S	Q	R	E	W	E
D	B	P	K	O	T	L	U	E	D
R	C	C	L	N	E	N	G	L	D
L	U	G	X	A	W	K	U	A	E
B	W	Z	S	K	P	P	B	O	R
I	X	H	O	V	C	D	V	F	M
A	P	P	L	E	S	R	C	S	L

SPEED
FOAL
BREED
LEASH
APPLES
SUGAR
MOUNT
BUCKSKIN

1. Compare and find by how many more or less the numbers are different.

68	>	33	by	35	27	<	95	by	68
14	<	51	by	37	80	>	45	by	35
65	>	17	by	48	28	<	77	by	49
66	<	96	by	30	91	>	18	by	73

2. Write the missing numbers near the letters to make the equations true.

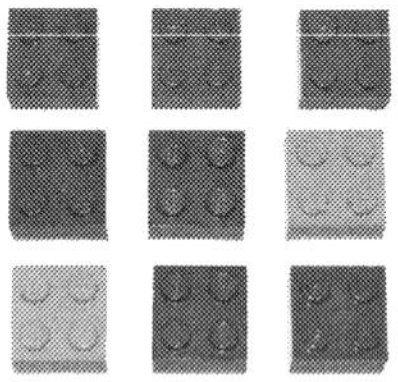

G 8 + G 8 + G 8 = 24

G 8 + G 8 + Y 15 = 31

Y 15 - B 1 + G 8 = 22

3. Find the value.

739586: tens + ten thousands – ones + hundreds + hundred thousands – thousands = 8+3-6+5+7-9=8.

Let's go through rounding one more time, eh? I don't get it. Lets' take 242 and round it to the nearest hundred. 242 may be rounded up to 300 or down to 200. Which one is correct?

Let's look at the range from 200 to 300. I will write 2 and 3 to show 2 hundreds (200) and 3 hundreds (300)

I write the numbers 0, 1, 2, 3, 4, 5, 6, 7, 8, 9 to show tens.

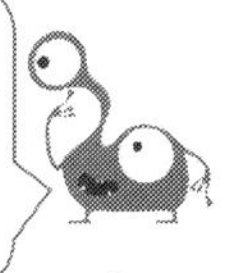

2 0 1 2 3 4 5 6 7 8 9 3

200 ← → 300

There are 10 digits and we divide them equally into 2 parts: 0, 1, 2, 3, 4. Then, we separate them from 5, 6, 7, 8, 9.

0, 1, 2, 3, 4 are close to 2 hundreds. So, when rounding, we go down and the numbers 208, 215, 227, 239, 246 are rounded down to 200 (look only at the next digit to the right – I underline the tens if I need to round to the nearest hundred).

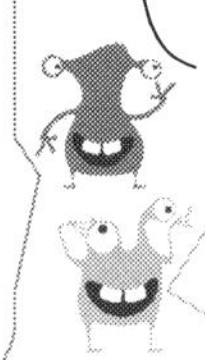

5, 6, 7, 8, 9 are close to 3 hundreds.

So, when rounding, we go up and the numbers 259, 266, 278, 284, 297 are rounded up to 300.

1. Write the missing numbers and an equation. Add. Use Round-Up-Strategy.

39 + 26 → 30, 4

tens	ones
1	
3	9
+ 2	6
6	5

74 + 17 → 20, 3

tens	ones
1	
7	4
+ 1	7
9	1

48 + 39 → 40, 1

tens	ones
1	
4	8
+ 3	9
8	7

39 + 26 = 39+30-4=65 74 + 17 = 74+20-3=91 48 + 39 = 48+40-1=87

2. Subtract.

64: 24 – 4, 28, 36

93: 63 – 3, 66, 27

58: 28 – 1, 29, 29

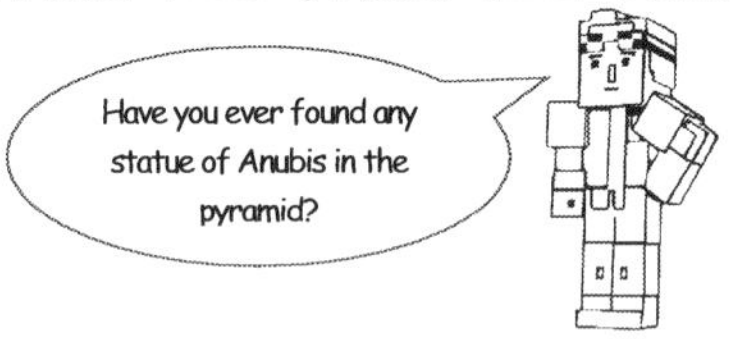

1. Fill in the missing "+", "-", or "=" to make the equations true. You can use the same sign ("+", "-", or "=") more than once.

12 = 31 + 43 - 62

96 - 85 + 18 = 29

25 + 32 - 21 = 36

0 = 96 - 64 - 32

76 + 21 - 42 = 55

28 = 81 - 65 + 12

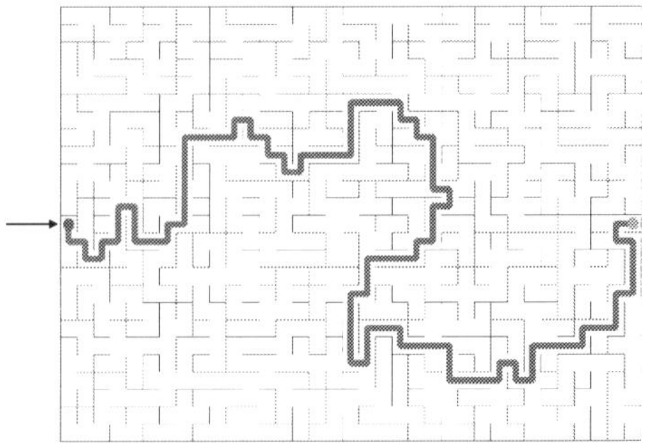

1. Subtract. Hint: 500 – 290 = 500 – 200 – 90 = 210.

800 – 260 = 540 400 – 170 = 230 300 – 150 = 150

900 – 530 = 370 600 – 280 = 320 700 – 640 = 60

2. Write the number words.

509 Five hundred nine

____

3. Subtract.

775 - 525 = 250 750 - 325 = 425 925 - 575 = 350

975 - 425 = 550 450 - 125 = 325 625 - 475 = 150

4. Round the numbers to the nearest hundred.

845 is 800	684 is 700	876 is 900
226 is 200	549 is 500	723 is 700
987 is 1000	669 is 700	184 is 200
116 is 100	358 is 400	356 is 400
924 is 900	721 is 700	157 is 200
467 is 500	473 is 500	491 is 500

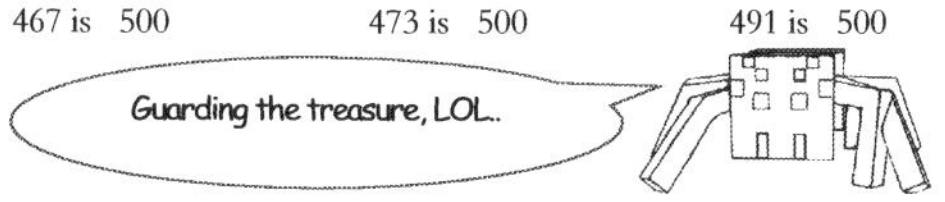

1. Write the missing numbers and an equation. Add or subtract. Use Round-Up-Strategy.

68 + 14	82 - 37	59 + 16	72 - 36
20 6	40 3	20 4	40 4

	tens	ones		tens	ones		tens	ones		tens	ones
	1			7	12		1			6	12
	6	8		~~8~~	~~2~~		5	9		~~7~~	~~2~~
+	1	4	-	3	7	+	1	6	-	3	6
	8	2		4	5		7	5		3	6

68+14 = 68+20-6=82 82 – 37 = 82-40+3=45 59 + 16 = 59+20-4=75 72 – 36 = 72-40+6=36

2. Subtract.

720 - 280 = 440 910 - 430 = 480 850 - 290 = 560

630 - 270 = 360 740 - 80 = 660 440 - 380 = 60

750 - 150 - 250 = 350 900 - 150 - 350 = 400

800 - 250 - 300 = 250 700 - 250 - 300 = 150

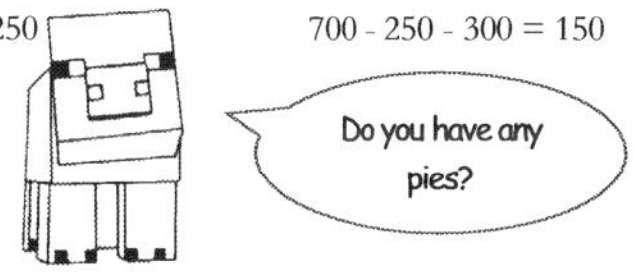

1. Write the missing numbers and an equation. Subtract or add.

48 + 23	65 - 49	57 + 18	84 - 45
2 21	45 4	3 15	44 1

	tens	ones		tens	ones		tens	ones		tens	ones
	1			5	15		1			7	14
	4	8		~~6~~	~~5~~		5	7		~~8~~	~~4~~
+	2	3	-	4	9	+	1	8	-	4	5
	7	1		1	6		7	5		3	9

48+23 = 48+2+21=71 65–49 = 65-45-4=16 57+18 = 57+3+15=75 84–45 = 84-44-1=39

2. Fill in the missing numbers to make the equations true.

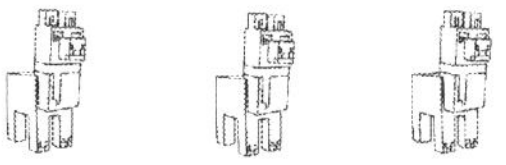

15 + 15 + 15 = 45

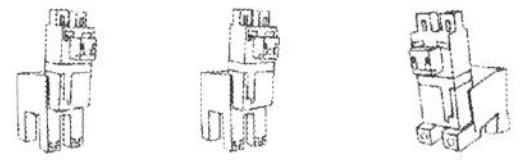

15 + 15 + 9 = 39

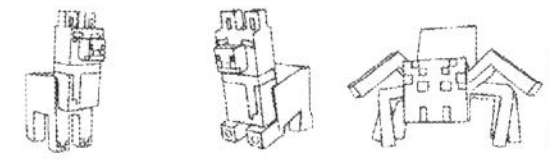

15 + 9 + 8 = 32

1. I have a favorite two-digit number:

- the sum of tens (T) and ones (O) is 15;
- their difference is 3;
- the ones digit is an even number.

What number is my favorite? 9 6.

T + O = 15 T - O = 3 O = even 9 6 ?

2. Continue a series of numbers.

3, 6, 12, 15, 21, 24, 30, 33, 39, 42, 48.

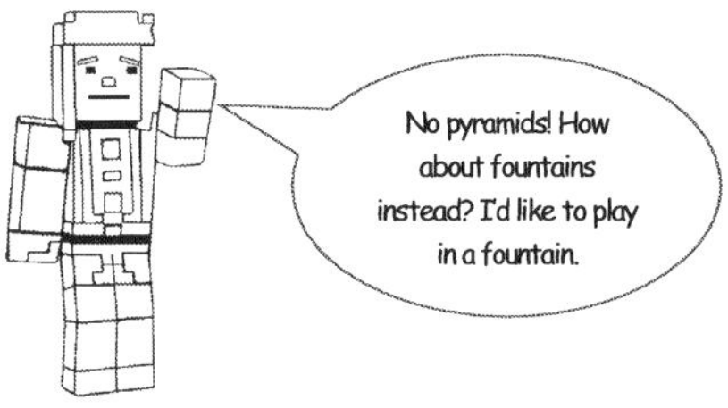

3. Fill in the missing signs "+" or/and "-" to make the equations true.

5 + 5 + 5 + 5 + 5 = 25 5 + 5 + 5 - 5 - 5 = 5

5 + 5 - 5 + 5 + 5 = 15 5 + 5 + 5 + 5 + 5 - 5 = 20

1. Write the missing numbers and an equation. Subtract.

76 - 28 → 26, 2

	tens	ones
	6	16
	~~7~~	~~6~~
-	2	8
	4	8

96 - 49 → 46, 3

	tens	ones
	8	16
	~~9~~	~~6~~
-	4	9
	4	7

57 - 38 → 37, 1

	tens	ones
	4	17
	~~5~~	~~7~~
-	3	8
	1	9

76 - 28 = 76-26-2=48   16 - 9 = 96-46-3=47   57 - 38 = 57-37-1=19

2. Subtract.

546 - 108 = 438   630 - 128 = 502   180 - 123 = 57

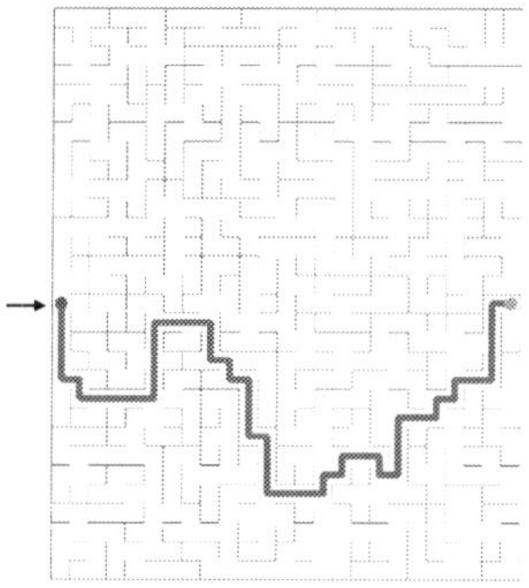

1. Unscramble each word to find the color of the horses in Minecraft . Write the word in the squares.

tsechunt	c	h	e	s	t	n	u	t
hewti	w	h	i	t	e			
calkb	b	l	a	c	k			
agyr	g	r	a	y				

1. Fill in the missing numbers so that the sums in the columns, rows, and two numeral diagonals are the same. Draw the arrows from the least number to the next greater number, and then to the greatest. Do you see any pattern? 7-14-21-28-35-42-49-56-63.

56	21	28
7	35	63
42	49	14

8	28	24
36	20	4
16	12	32

2. Subtract.

450 - 162 = 288   975 - 338 = 637   715 - 242 = 473

493 - 364 = 129   526 - 370 = 156   763 - 274 = 489

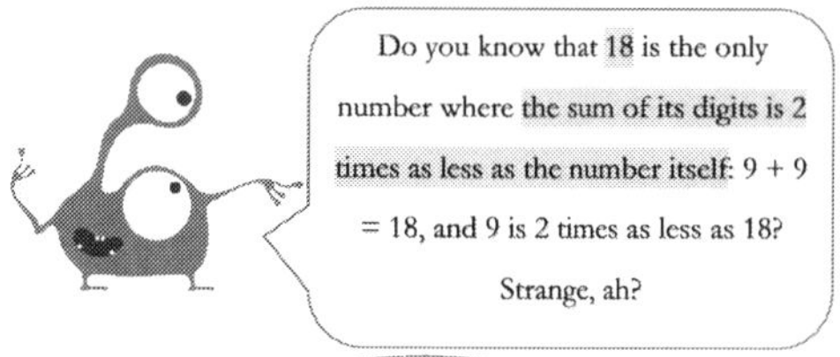

1. Write the missing numbers and an equation. Subtract or add. Follow the strategy.

58 + 25 → 2, 23

	tens	ones
	1	
	5	8
+	2	5
	8	3

22 - 13 → 12, 1

	tens	ones
	1	12
	~~2~~	~~2~~
-	1	3
		9

17 + 16 → 3, 13

	tens	ones
	1	
	1	7
+	1	6
	3	3

34 - 26 → 24, 2

	tens	ones
	2	14
	~~3~~	~~4~~
-	2	6
		8

58 + 25 = 58+2+23=83   22 – 13 = 22-12-1=9   17 + 16 = 17+3+13=33   34 – 26 = 34-24-2=8

4. Round the numbers to the nearest thousand.

2845 is 3000	6184 is 6000	8756 is 9000
2526 is 3000	5549 is 6000	2723 is 3000
9187 is 9000	9669 is 10000	1184 is 1000
1716 is 2000	3528 is 4000	3256 is 3000
9254 is 9000	7121 is 7000	1657 is 2000
4607 is 5000	4873 is 5000	4951 is 5000

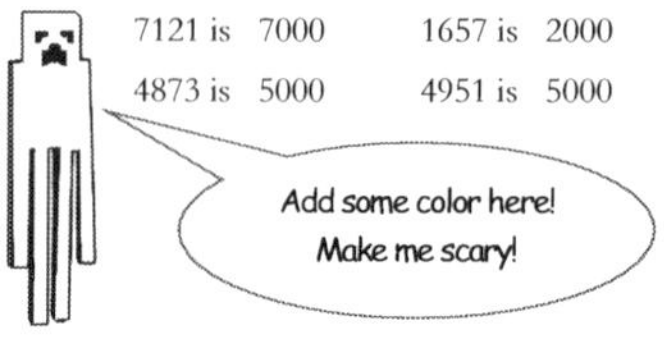

1. Write the missing numbers and an equation. Subtract or add. Use Round-Up-Strategy or decompose.

76 + 18	61 - 36	69 + 25	52 - 29
20 2	40 4	1 24	22 7

tens	ones		tens	ones		tens	ones		tens	ones
1			5	11		1			4	12
7	6		~~6~~	~~1~~		6	9		~~5~~	~~2~~
+ 1	8		- 3	6		+ 2	5		- 2	9
9	4		2	5		9	4		2	3

76 + 18 = 76+20-2=94

61 – 36 = 61-40+4=25

69 + 25 = 69+1+24=94

52 – 29 = 52-22-7=23

2. Round the numbers to the nearest ten.

2845 is 2850	6184 is 6180	8756 is 8760
2526 is 2530	5549 is 5550	2723 is 2720
9187 is 9190	9669 is 9670	1184 is 1180
1716 is 1720	3528 is 3530	3256 is 3260
9254 is 9250	7121 is 7120	1657 is 1660
4607 is 4610	4873 is 4870	4951 is 4950

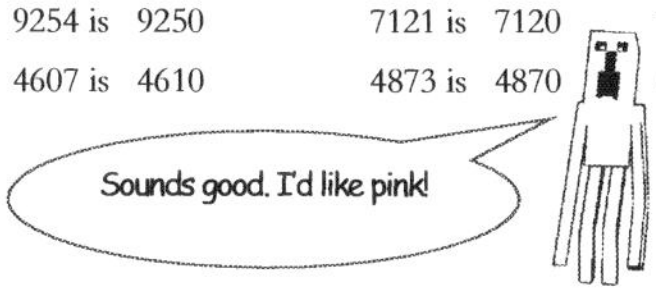

1. Write the missing numbers and an equation. Subtract or add. Follow the strategy.

74 + 19	71 - 35	47 + 35	56 - 27
6 13	31 4	40 5	30 3

tens	ones		tens	ones		tens	ones		tens	ones
1			6	11		1			4	16
7	4		~~7~~	~~1~~		4	7		~~5~~	~~6~~
+ 1	9		- 3	5		+ 3	5		- 2	7
9	3		3	6		8	2		2	9

74 + 19 = 74+3+13=93

71 – 65 = 71-31-4=36

47 + 35 = 47-40+5=82

56 – 37 = 56-30+3=29

2. Subtract and add. Use Round-Up-Strategy or decompose.

546 - 108 = 438	630 - 128 = 502	180 - 123 = 57
840 - 245 = 595	562 - 158 = 404	870 - 124 = 746
987 - 351 = 636	275 - 269 = 6	753 - 225 = 528
550 + 80 = 630	273 + 40 = 313	475 + 50 = 525
161 + 70 = 231	510 + 85 = 595	920 + 94 = 1014

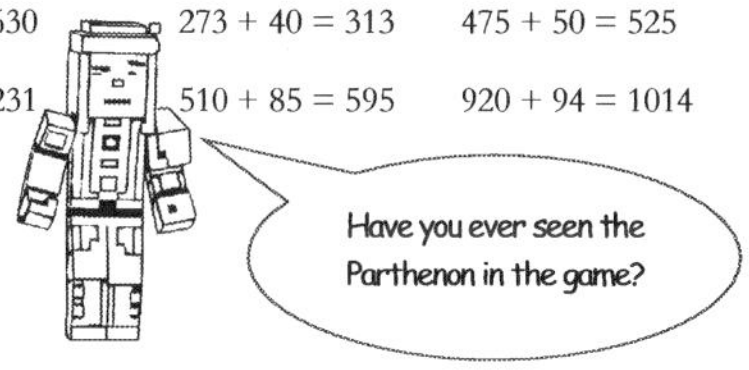

1. Subtract.

64	37	53	95	88
- 31	- 15	- 41	- 74	- 27
33	22	12	21	61

80	40	60	70	30
- 6	- 7	- 2	- 5	- 9
74	33	58	65	21

80	40	70	60	30
- 36	- 24	- 52	- 15	- 25
44	16	18	45	5

1. Write the missing numbers 12, 16, 18, 31, and 45 to make the equation true.

45 - 18 = 31 - 16 + 12

2. Answer the question.

A	B	C	D	E	F	G	H	I	J	K	L	M
1	2	3	4	5	6	7	8	9	10	11	12	13
N	O	P	Q	R	S	T	U	V	W	X	Y	Z
14	15	16	17	18	19	20	21	22	23	24	25	26

Who is the largest mob in the game?
23 8 15 / 9 19 / 20 8 5 / 12 1 18 7 5 19 20 / 13 15 2 / 9 14 / 20 8 5 / 7 1 13 5

The Ender dragon
20 18 5 / 5 14 4 5 18 / 4 18 1 7 15 14

1. Write the missing numbers and an equation. Subtract or add.

60 - 29	50 - 34	70 - 46
30 1	40 6	50 4

tens	ones		tens	ones		tens	ones
5	10		4	10		6	10
~~6~~	~~0~~		~~5~~	~~0~~		~~7~~	~~0~~
- 2	9		- 3	4		- 4	6
3	1		1	6		2	4

60 – 29 = 60-30+1=31

50 – 34 = 50-40+6=16

70 – 46 = 70-50+4=24

2. Add.

300 + 400 + 150 = 850

350 + 450 + 100 = 900

750 + 150 + 100 = 1000

650 + 175 + 125 = 950

550 + 150 + 300 = 1000

450 + 250 + 300 = 1000

3. Subtract. Use Round-Up-Strategy.

850 – 80 = 770

250 – 90 = 160

585 – 90 = 495

974 – 80 = 894

403 – 50 = 353

109 – 50 = 59

2. Find and circle or cross out the words to find more about lighting in Minecraft.

L	P	T	A	T	R	Y	T	Z	B
I	X	F	H	F	O	R	U	R	Z
G	W	K	N	G	X	R	I	Y	K
H	T	E	R	N	I	G	C	Q	E
T	S	T	E	S	H	L	C	H	T
G	Q	H	T	T	U	V	N	E	B
S	S	E	N	K	R	A	D	U	C
X	A	E	A	M	L	Q	Y	X	S
Y	S	U	L	K	R	L	L	N	H
S	F	U	R	N	A	C	E	S	J

DARKNESS
SUNLIGHT
TORCH
LANTERN
BRIGHTNESS
FURNACE
LIGHT

1. Write the missing numbers and an equation. Subtract or add.

59 + 35	62 - 14	66 + 28	94 - 68
1 34	12 2	4 24	64 4

tens	ones		tens	ones		tens	ones		tens	ones
1			5	12		1			8	14
5	9		~~6~~	~~2~~		6	6		~~9~~	4
+ 3	5		- 1	4		+ 2	8		- 6	8
9	4		4	8		9	4		2	6

59 + 35 = 59+1+34=94

62 – 14 = 62-12-2=48

66 + 28 = 66+4+24=94

94 – 68 = 94-64-4=26

2. Round the numbers to the nearest hundred.

2851 is 2900	6192 is 6200	5156 is 5200
2528 is 2500	5565 is 5600	9323 is 9300
9134 is 9100	9637 is 9600	4284 is 4300
1781 is 1800	3554 is 3600	7356 is 7400
9276 is 9300	7197 is 7200	4857 is 4900
4612 is 4600	4861 is 4900	9551 is 9600

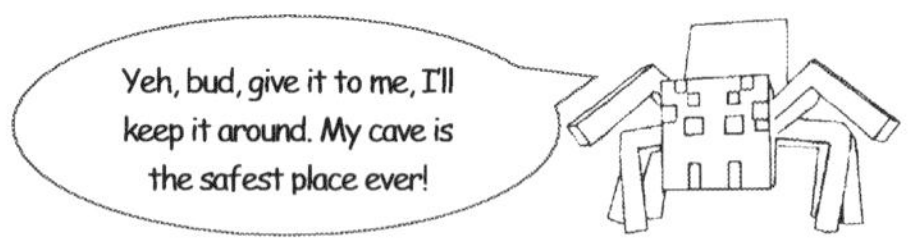

1. Write the missing numbers and an equation. Subtract or add. Use Round-Up-Strategy.

77 + 15	48 - 19	18 + 67	57 - 29
20 5	20 1	20 2	30 1

tens	ones		tens	ones		tens	ones		tens	ones
1			3	18		1			4	17
7	7		4	~~8~~		6	7		~~5~~	~~7~~
+ 1	5		- 1	9		+ 1	8		- 2	9
9	2		2	9		8	5		2	8

77 + 15 = 77+20-5=92

48 – 19 = 48-20+1=29

18 + 67 = 67+20-2=85

57 – 29 = 57-30+1=28

2. Add.

250 + 175 + 350 = 775

425 + 250 + 375 = 1050

325 + 225 + 475 = 1025

625 + 150 + 255 = 1030

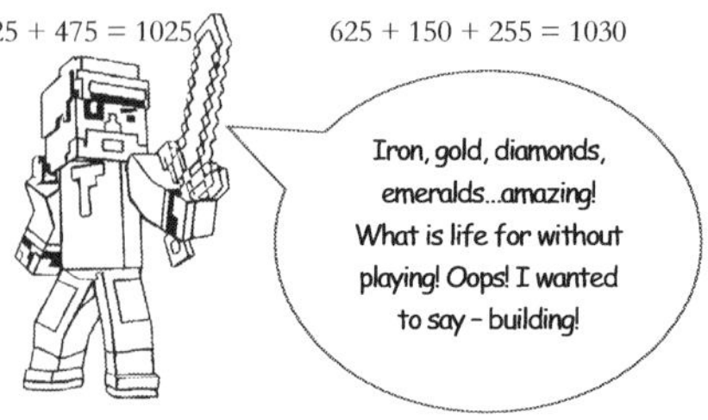

Hint: Ones under ones, tens under tens, hundreds under hundreds;
Subtract ones, then tens, then hundreds.

400 - 76

70 6

Decompose 76. 76 is 70+6.

	hundreds	tens	ones
	4	0	0
-		7	6
	__	__	__

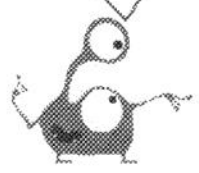

I cannot subtract 6 ones from 0 ones: 0 – 6 = ?!? Or 7 tens from 0 tens. You must get more ones and more tens.

400-76=400-70-6=330-6=324 or 400-76=400-80+4=320+4=324. I solved it! High five!

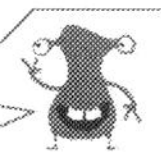

Step 1: I subtract in columns, I need one more row above 400.

Step 2: Borrow 1 ten of the 40 tens. Write 10 ones above 0 ones since 1 ten equals 10 ones. Write 39 above 40 tens: 40-1=39.

Step 3: Cross out 0 ones and 40 tens.

	hundreds	tens	ones
	3	9	10
	4	~~0~~	~~0~~
		7	6
	__	__	__

Step 4: Subtract 6 ones from 10 ones: 10-6=4.

Step 5: Subtract 7 tens from 9 tens:9-7=2.

Step 6: Rewrite 3 since 3 – 0 = 3.

	hundreds	tens	ones
	3	9	10
	4	~~0~~	~~0~~
-		7	6
	3	2	4

1. Subtract. Fill in the missing numbers.

96-41=96-40-1

35	6	55

93-38=93-40+2

11	27	55

72-44=72-42-2

19	25	28

2. Subtract.

600 - 71 → 70, 1

	Hs	Ts	Os
	5	9	10
	~~6~~	~~0~~	~~0~~
-		7	1
	5	2	9

600 - 71 = 600-70-1=529

500 - 48 → 40, 8

	Hs	Ts	Os
	4	9	10
	~~5~~	~~0~~	~~0~~
-		4	8
	4	5	2

500 – 48 = 500-40-8=452

300 - 66 → 60, 6

	Hs	Ts	Os
	2	9	10
	~~3~~	~~0~~	~~0~~
-		6	6
	2	3	4

300 - 66 = 300-60-6=234

1. Fill in the missing whole numbers.

350 → 178, 172; 178 → 89, 89; 172 → 67, 105

301 → 184, 117; 184 → 105, 79; 117 → 59, 58

2. Unscramble each word to find out more about the Nether. Write the word in the squares.

tresofrs	f o r t r e s s
gmmaa	m a g m a
idrgeb	b r i d g e
inopot	p o t i o n
toparl	p o r t a l

1. Add.

492	847	426	722
+ 531	+ 225	+ 158	+ 536
1 023	1 072	584	1 258
686	564	252	296
+ 153	+ 333	+ 175	+ 188
839	897	427	484

2. Round the numbers to the nearest hundred.

3497 is 3500	1025 is 1000	5402 is 5400
8214 is 8200	7943 is 7900	7605 is 7600
9734 is 9700	3059 is 3100	9124 is 9100
8246 is 8200	4167 is 4200	5426 is 5400
9214 is 9200	9802 is 9800	1105 is 1100
7635 is 7600	7319 is 7300	7631 is 7600

1. Subtract. Follow the strategy.

500 - 82 → 90, 8

Hs	Ts	Os
4	9	10
~~5~~	~~0~~	~~0~~
-	8	2
4	1	8

700 - 37 → 40, 3

Hs	Ts	Os
6	9	10
~~7~~	~~0~~	~~0~~
-	3	7
6	6	3

400 - 65 → 70, 5

Hs	Ts	Os
3	9	10
~~4~~	~~0~~	~~0~~
-	6	5
3	3	5

500 - 82 = 500-90+8=418 700 – 37 = 700-40+3=663 400 - 65 = 400-70+5=335

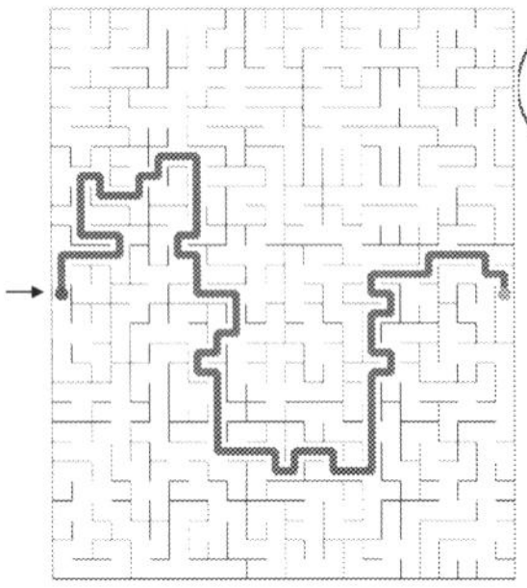

CHAPTER 2

Fun with Renaming/Regrouping

Adding 3-Digit Numbers Vertically up to 1000
Adding 3-Digit Numbers Horizontally up to 1000
Subtracting 3-Digit Numbers Vertically
Subtracting 3-Digit Numbers Horizontally
Adding 3-Digit Numbers Vertically beyond 1000
Adding 3-Digit Numbers Horizontally beyond 1000
Subtracting 4-Digit Numbers Horizontally
Number Talks
Olympiad Math Problems
CogAT test prep
Puzzles
Mazes
Word Search
Cryptograms

1. Write the missing numbers and an equation. Hint: use Round-Up Strategy.

700 - 37 → 40, 3

Hs	Ts	Os
6	9	10
~~7~~	~~0~~	~~0~~
-	3	7
6	6	3

900 - 58 → 60, 2

Hs	Ts	Os
8	9	10
~~9~~	~~0~~	~~0~~
-	5	8
8	4	2

400 - 83 → 90, 7

Hs	Ts	Os
3	9	10
~~4~~	~~0~~	~~0~~
-	8	3
3	1	7

700 - 37 = 700-40+3=663 900 – 58 = 900-60+2=842 400 - 83 = 400-90+7=317

2. Subtract or add.

362 – 8 = 354	874 – 9 = 865	965 – 7 = 958
84 + 37 = 121	95 + 68 = 163	76 + 69 = 145
700 + 240 = 940	600 + 330 = 930	300 + 480 = 780
400 – 225 = 175	300 – 175 = 125	800 – 455 = 345
500 – 285 = 215	900 – 745 = 155	600 – 395 = 205

1. Write the missing numbers and an equation. Hint use Round-Up Strategy.

400 - 35 → 40, 5

Hs	Ts	Os
3	9	10
4	~~0~~	~~0~~
-	3	5
3	6	5

600 - 94 → 100, 6

Hs	Ts	Os
5	9	10
~~6~~	~~0~~	~~0~~
-	9	4
5	0	6

900 - 51 → 60, 1

Hs	Ts	Os
8	9	10
~~9~~	~~0~~	~~0~~
-	5	1
8	4	9

400 - 35 = 400-40+5=365 600 – 94 = 600-100+6=506 900 - 51 = 900-60+9=849

2. Fill in the missing numbers.

	6	3
-	3	7
	2	6

Hm…think what number minus 7 gives us 6? 13, write 3 in the top row. Remember 1 because you subtract 7 out of 3. Subtract 6 – 1 - 2 to get the number in the second row. Got it?

3. Continue a series of numbers:

61, 74, 87, 100, 113, 126. 139, 152, 165

1. Do you like a challenge? I DO! Let's take a timer and see who will solve these problems faster. Adults are welcome, too!

Name ______________________ Time ____ : ____.

9 + 2 = 11	8 + 3 = 11	7 + 4 = 11	6 + 5 = 11
9 + 3 = 12	8 + 4 = 12	7 + 5 = 12	6 + 6 = 12
9 + 4 = 13	8 + 5 = 13	7 + 6 = 13	6 + 7 = 13
9 + 5 = 14	8 + 6 = 14	7 + 7 = 14	6 + 8 = 14
9 + 6 = 15	8 + 7 = 15	7 + 8 = 15	6 + 9 = 15
9 + 7 = 16	8 + 8 = 16	7 + 9 = 16	9 + 8 = 17
8 + 9 = 17	9 + 9 = 18	18– 9 = 9	17 – 8 = 9
17 – 9 = 8	16 – 7 = 9	16 – 8 = 8	16 – 9 = 7
15 – 6 = 9	15 – 7 = 8	15 – 8 = 7	15 – 9 = 6
14 – 5 = 9	14 – 6 = 8	14 – 7 = 7	14 – 8 = 6
14 – 9 = 5	13 – 4 = 9	13 – 5 = 8	13 – 6 = 7
13 – 7 = 6	13 – 8 = 5	13 – 9 = 4	12 – 3 = 9
12 – 4 = 8	12 – 5 = 7	12 – 6 = 6	12 – 7 = 5
12 – 8 = 4	12 – 9 = 3	11 – 2 = 9	11 – 3 = 8
11 – 4 = 7	11 – 5 = 6	11 – 6 = 5	11 –7 = 4

2. Continue a series of numbers:

909, 808, 707, 606, 505, 404, 303, 202

About the Greek times. Do you know the difference between "ingots" and "helots"?

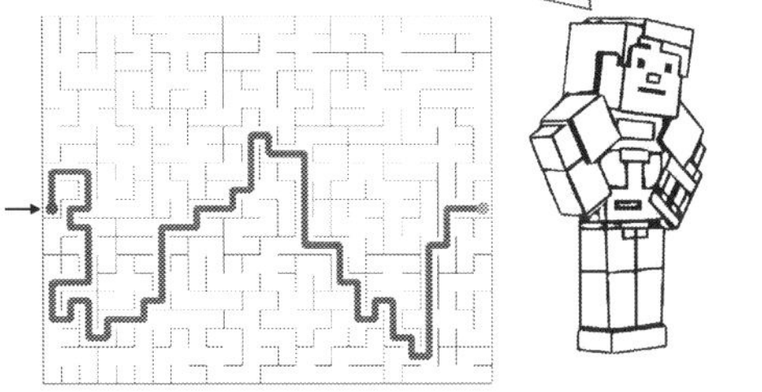

1. Find and circle or cross out the words to find out more about the Ender Dragon.

C	R	Y	S	T	A	L	S	S	H
U	W	D	P	M	P	I	L	G	E
I	Z	P	E	U	R	L	N	G	A
I	K	I	R	S	A	L	V	E	L
M	F	P	G	B	T	G	O	V	I
L	L	C	W	D	V	R	P	G	N
E	G	O	C	Y	M	B	O	P	G
D	N	Y	K	C	I	R	T	Y	H
S	S	O	T	Z	F	N	T	U	F
J	R	A	S	Y	W	N	C	O	L

CRYSTALS
PURPLE
DESTROY
SNOWBALLS
EGGS
TRICKY
GOLEMS
HEALING

Hint: Write ones under ones, tens under tens, and so on; Add ones first, then tens, then hundreds; Carry when the sum in any column is 10 or more.

526 + 175

4 171

hund-reds	tens	ones
5	2	6
+ 1	7	5
—	—	—

Decompose 175. 175 is 4+171.

6+5=11. So, you write 1 in one's place and carry 1 ten with the other tens.

Aha, I need to add one more row to write the numbers that were carried, right?

hund-reds	tens	ones
—	—	
5	2	6
+ 1	7	5
—	—	—

Step 1: I need one more row above 526.

Step 2: First, I add ones: 6+5=11.

Write the 1 in one's place. Carry 1 ten with the other tens. Write the 1 above 2 in ten's place.

hund-reds	tens	ones
—	1	
5	2	6
+ 1	7	5
—	—	1

Step 3: Add tens: 1+2+7=10. Write the 0 in ten's place. Carry 1 hundred with the other hundreds. Write the 1 above 5 in hundred's place.

Step 4. Add hundreds: 1+5+1=7.

hund-reds	tens	ones
1	1	
5	2	5
+ 1	7	5
7	0	1

1. Write the missing numbers and an equation. Decompose.

397 + 65 → 3, 62

	Hs	Ts	Os
	1		
	3	9	7
+		6	5
	4	6	2

397 + 65 = 397+3+62=462

481 - 79 → 71, 8

	Hs	Ts	Os
		7	11
	4	~~8~~	~~1~~
-		7	9
	4	0	2

481 – 79 = 481-71-8=402

382 + 58 → 18, 40

	Hs	Ts	Os
	1	1	
	3	8	2
+		5	8
	4	4	0

382 + 58 = 382+18+40=440

2. Round the numbers to the nearest hundred.

451	500	516	500	623	600	782	800	843	800	994	1000
427	400	569	600	657	700	796	800	865	900	938	900
494	500	584	600	615	600	744	700	811	700	962	1000
448	400	507	500	608	600	729	700	888	900	927	900

3. Continue a series of numbers:

4, 6, 10, 16, 24, 34, 46, 60, 76, 94

1. Evaluate each expression.

28 + 7 – 15 = 20 37 – 9 + 22 = 50 66 + 5 – 23 = 48

82 – 6 + 14 = 90 54 + 8 – 17 = 45 74 – 7 + 11 = 75

42 + 8 – 36 = 14 70 – 6 + 16 = 80 62 + 9 – 43 = 28

2. Add.

Hint: 250+650 = 650+50+200 or 200+600+50+50 or 250+50+600.

150 + 450 = 600 350 + 350 = 700 450 + 550 = 1000

150 + 850 = 1000 750 + 50 = 800 350 + 650 = 1000

550 + 150 = 700 850 + 50 = 900 350 + 450 = 800

450 + 450 = 900 250 + 150 = 400 150 + 750 = 900

250 + 550 = 800 450 + 250 = 700 650 + 50 = 700

3. Find the value.

99: 45 | 38 | 16

78: 12 | 27 | 39

88: 44 | 19 | 25

1. Subtract.

500	800	507	592
- 225	- 635	- 369	- 386
275	165	138	206
300	400	1000	1000
- 175	- 385	- 565	- 397
125	15	435	603

2. Write the missing numbers to make the equations true.

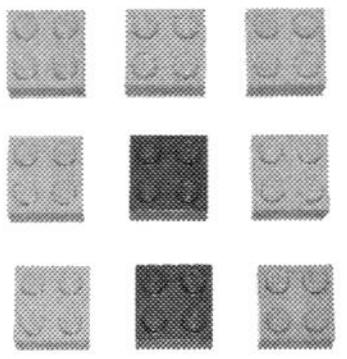

Y12 + Y 12 + Y 12 = 36

Y 12 + B 9 + Y 12 = 33

Y 12 - B 9 + G 19 = 22

3. Fill in the missing numbers.

4 7	5 2	8 3
+ 4 9	+ 2 8	- 2 6
9 6	8 0	5 7

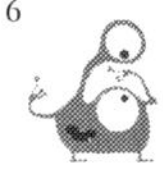

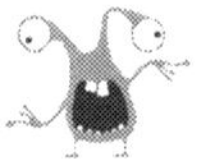

1. Subtract. Hint: use Round-Up-Strategy.

500 – 54 = 446 200 – 68 = 132 900 – 83 = 817

400 – 16 = 384 700 – 39 = 661 600 – 42 = 558

2. Add. Hint: decompose.

14 + 65 = 79 26 + 34 = 60 28 + 59 = 87

57 + 25 = 82 27 + 64 = 91 17 + 35 = 52

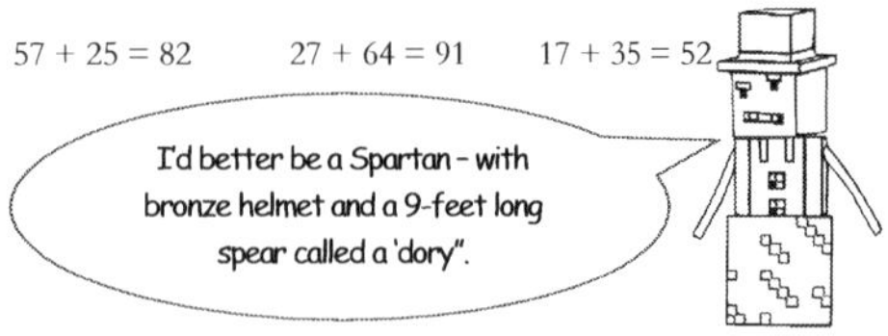

3. Subtract.

245 - 176 → 145, 31

Hs	Ts	Os
1	13	15
~~2~~	~~4~~	~~5~~
- 1	7	6
	6	9

364 - 219 → 214, 5

Hs	Ts	Os
	5	14
3	6	4
- 2	1	9
1	4	5

853 - 466 → 453, 13

Hs	Ts	Os
7	14	13
8	5	3
- 4	6	6
3	8	7

245 – 176 = 245 – 145 – 31 = 69 364 – 219 = 364- 214-5=145 853 – 466 = 853- 453-13=387

1. Evaluate each equation. Use the alphabet on page "2" of the workbook to find out the hidden word.

100 - 47 - 35 + 28 – 20 = 26

100 - 53 - 16 + 36 - 58 = 9

100 - 39 - 13 – 23 - 18 = 7

100 - 29 - 38 + 13 - 20 = 26

100 - 45 - 24 + 11 – 41 = 1

100 - 19 - 28 - 36 – 10 = 7

26	9	7	26	1	7
Z	I	G	Z	A	G

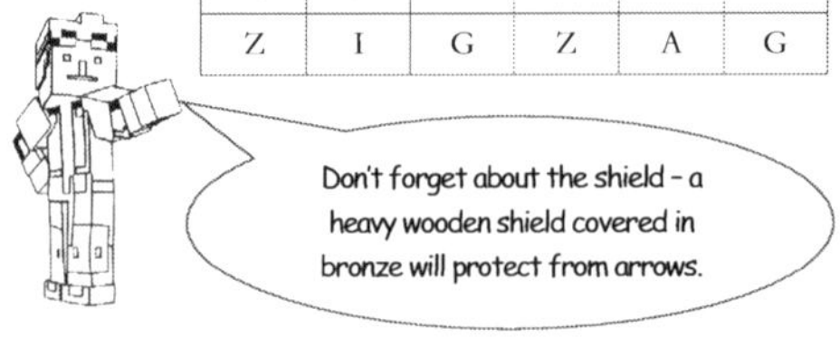

2. Subtract.

300 – 78 = 222 600 – 36 = 564 800 – 59 = 741

3. Continue a series of numbers:

4, 4, 8, 12, 20, 32, 52, 84, 136, 220

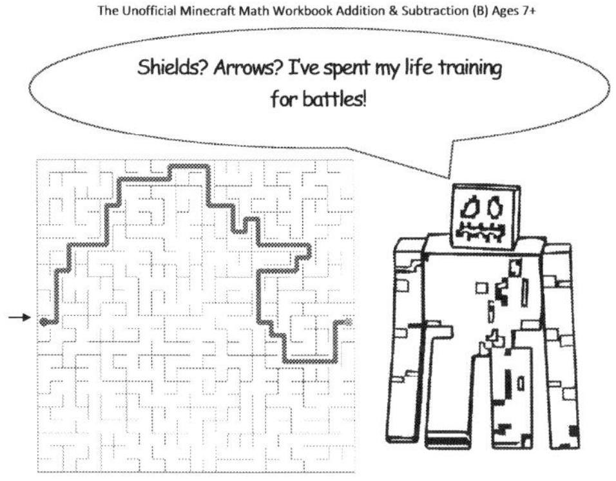

1. Unscramble each word to find out more about the fireworks in Minecraft. Write the word in the squares.

eexdlop	e	x	p	l	o	d	e	
gbna	b	a	n	g				
ewinklt	t	w	i	n	k	l	e	
etocrk	r	o	c	k	e	t		
uwroegdnp	g	u	n	p	o	w	d	e	r

1. Fill in the numbers 20, 22, 24, 26, 28, 30, 32, 34, 36 into the squares so that the sum in each row and in each column is 84.

Draw the arrows from the least number to a greater number up to the greatest number: 20 - 22 - 24, etc.

Answer:

20-22-24-26-28-30-32-34-36

34	24	26
20	28	36
30	32	22

2. Continue a series of numbers:

225, 275, 250, 300, 275, 325, 300, 350, 325, 375, 350

3. Add.

6	7	5	8
10	80	90	50
+ 267	+ 184	+ 232	+ 294
283	271	327	352

4	3	9	7
26	87	74	74
+ 615	+ 429	+ 526	+ 385
645	519	609	466

1. Solve the pattern and find the missing number.

16 15
53
13 9

14 37
20
19 12

17 35
11
23 18

15 27
83
12 29

2. Fill in the missing numbers and answer the question.

I have 71 bricks. How many more bricks could I add to change the tens by 1 digit up? Find as many options as you can.

9, 10, 11, 12, 13, 14, 15, 16, 17,
18.

3. Continue a series of numbers:

15, 30, 45, 60, 75, 90, 105, 120.

1. Add.

496 + 158
4 154

Hs	Ts	Os
1	1	
4	9	6
+ 1	5	8
6	5	4

496 + 158 = 654

459 + 377
41 336

Hs	Ts	Os
1	1	
4	5	9
+ 3	7	7
8	3	6

459 + 377 = 836

584 + 149
16 133

Hs	Ts	Os
1	1	
5	8	4
+ 1	4	9
7	3	3

584 + 149= 733

2. Round the numbers to the nearest hundred.

256	300	335	300	401	400	545	500	623	600	779	800
241	200	372	400	464	500	599	600	658	700	794	800
293	300	327	300	438	400	553	600	632	600	735	700
218	200	389	400	470	500	576	600	610	600	797	800

3. Continue a series of numbers:

63, 61, 58, 54, 49, 43, 36, 28, 19, 9

1. Write "+" or "-" to make the equations true.

56 + 19 - 27 = 48 34 - 17 + 25 = 42

100 - 64 + 18 = 54 69 - 38 - 24 = 7

2. Write the missing numbers and answer the question.

I have 80 bricks. How many bricks could I subtract to change the tens by 1 digit down? Find as many options as you can.

1, 2, 3, 4, 5, 6, 7, 8, 9, 10.

3. Add and compare the sums. Hint: 465 + 200 - 11.

465 + 189 = 654 > 314 + 295 = 609

257 + 497 = 754 < 164 + 298 = 462

4. Subtract.

1000 – 620 = 380 1000 – 490 = 510 1000 – 170 = 830

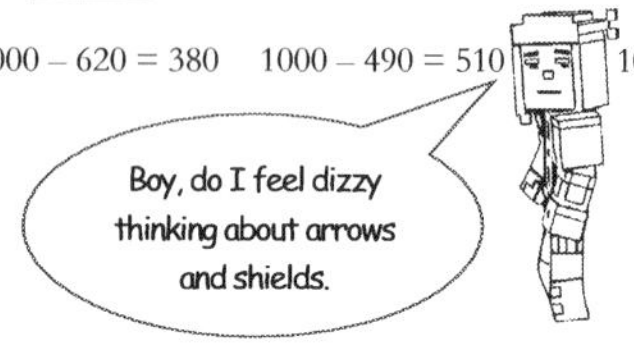

1. Write the missing numbers.

45 + ? 30 = 75 ? 55 + 5 = 60 85 – ? 47 = 38

24 + ? 41 = 65 ? 68 + 30 = 98 90 – ? 39 = 51

28 + ? 39 = 67 ? 27 + 22 = 49 65 – ? 47 = 18

24 + ? 21 = 45 ? 31 + 19 = 50 71 – ? 37 = 34

37 + ? 29 = 66 ? 42 + 12 = 54 57 – ? 28 = 29

48 + ? 41 = 89 ? 43 + 28 = 71 44 – ? 26 = 18

2. Write 3 different strategies to add and subtract: Decompose, Round-Up, Tens+Ones.

44+25+17= ____________=____________=86

65+14+29= ____________=____________=108

44+48–17= ____________=____________=75

65+17-29= ____________=____________=53

1. Add or subtract

563 + 128 → 130, 2

534 - 228 → 230, 2

391 + 199 → 200, 1

	Hs	Ts	Os
	1	1	
	5	6	3
+	1	2	8
	6	9	1

	Hs	Ts	Os
		2	14
	5	~~3~~	4
-	2	2	8
	3	0	6

	Hs	Ts	Os
	1	1	
	3	9	1
+	1	9	9
	5	9	0

563 + 128 = 691 534 – 228 = 306 391 + 199 = 590

2. Add. Write 2 strategies for each problem. The first one is done for you.

35 + 15 + 26 = 30 + 10 + 20 + 10 + 6 = 40 + 10 + 26 = 76

36 + 44 + 17 = ______________ = ______________ = 97

55 + 25 + 19 = ______________ = ______________ = 99

19 + 44 + 28 = ______________ = ______________ = 91

47 + 14 + 28 = ______________ = ______________ = 89

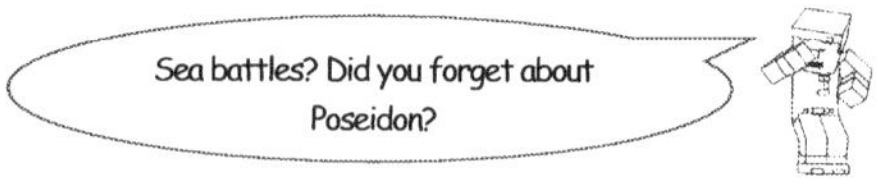

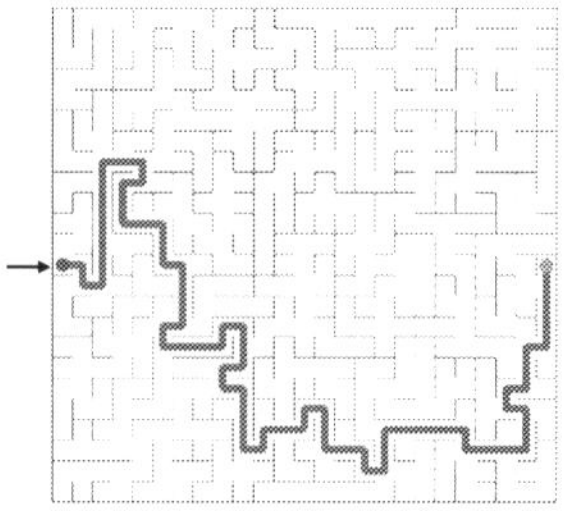

1. Find and circle or cross out the words to find out more about maps in Minecraft.

O	S	R	K	W	K	E	B	N	S
R	V	T	S	O	C	E	O	C	U
H	J	Z	A	W	M	I	O	C	R
A	D	V	E	N	T	U	R	E	V
N	M	K	U	A	D	E	C	P	I
F	M	Q	V	Y	A	A	U	P	V
M	P	E	O	T	X	J	R	F	A
Q	L	K	I	Z	A	A	M	D	L
E	O	V	E	M	A	R	F	Y	J
E	E	J	D	D	T	B	A	B	R

ELEVATION
STANDARD
ZOOM
FRAME
ADVENTURE
SURVIVAL
CREATIVE

1. Add. Follow the strategy.

239 + 263 → 300, 37

	Hs	Ts	Os
	1	1	
	2	3	9
+	2	6	3
	5	0	2

239 + 263 = 502

674 + 177 → 200, 23

	Hs	Ts	Os
	1	1	
	6	7	4
+	1	7	7
	8	5	1

674 + 177 = 851

448 + 396 → 400, 4

	Hs	Ts	Os
	1	1	
	4	4	8
+	3	9	6
	8	4	4

448 + 396 = 844

2. Add. Draw the arrows to show easier calculations.

91 + 55 + 22 + 15 + 19 + 48 =

= 250

56 + 26 + 23 + 14 + 17 + 44 =

= 180

24 + 27 + 21 + 23 + 26 + 29 =

= 150

3. Subtract.

1000 – 540 = 460 1000 – 360 = 640 1000 – 790 = 210

1. Evaluate each equation. Use the alphabet on page "2" of the workbook to find out the hidden word.

28 + 33 + 29 – 64 – 20 =6

23 + 59 + 2 – 37 - 16 = 1

14 + 18 + 29 – 28 – 21 =12

36 + 17 + 16 – 24 – 26 =19

52 + 17 + 19 – 22 – 41 =5

6	1	12	19	5
F	A	L	S	E

2. Write the missing numbers to make the sum and difference true.

		6	7
+		4	3
	1	1	0

	3	5	6
-		3	8
	3	1	8

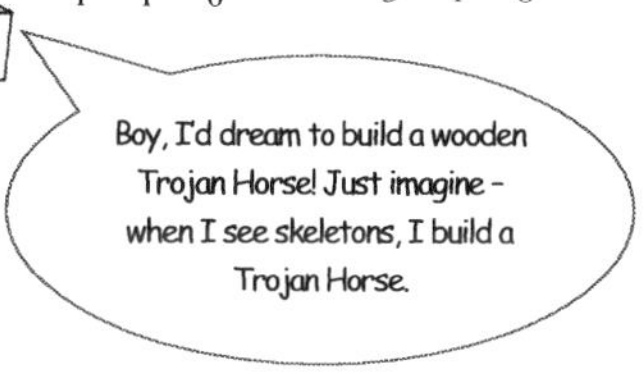

1. Evaluate each equation. Use the alphabet on page "2" of the workbook to find out the hidden word.

71 + 24 - 18 – 15 – 58 = 4

12 + 29 + 17 – 26 - 23 = 9

53 + 16 + 17 – 43 – 17 =26

41 + 15 + 18 – 36 – 12 =26

80 - 14 - 19 + 8 – 30 = 25

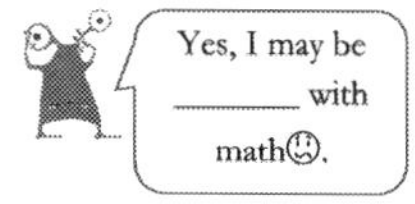

4	9	26	26	25
D	I	Z	Z	Y

2. Continue the pattern:

1999, 1888, 1666, 1333, 889, 334

3. Add and subtract.

800 – 458 = 342

135 + 835 = 970

700 – 282 = 418

1000 – 225 = 775

1000 – 328 = 672 1000 – 768 = 232

1. Subtract.

381 - 193 → 181, 12

	Hs	Ts	Os
	2	17	11
	~~3~~	~~8~~	~~1~~
-	1	9	3
	1	8	8

381 – 193 = 188

435 - 157 → 135, 22

	Hs	Ts	Os
	3	12	15
	~~4~~	~~3~~	~~5~~
-	1	5	7
	2	7	8

435 – 157 = 278

574 - 186 → 174, 12

	Hs	Ts	Os
	4	16	14
	~~5~~	~~7~~	4
-	1	8	6
	3	8	8

574 – 186 = 388

2. Add.

Hint: 154 + 67 = 154 + 100 - 33 or 154 + 6 + 60 + 1☺.

142 + 89 = 231

126 + 75 = 201

135 + 39 = 174

3. Write the missing numbers to make the sum and difference true.

	8	0	0
-	4	5	5
	3	4	5

	1	6	3
+	4	8	8
	6	5	1

1. Add.

189 + 146	198 + 374	265 + 247
11 135	2 372	35 212

Hs	Ts	Os		Hs	Ts	Os		Hs	Ts	Os
1	1			1	1			1	1	
1	8	9		1	9	8		2	6	5
+ 1	4	6		+ 3	7	4		+ 2	4	7
3	3	5		5	7	2		5	1	2

189 + 146 = 335   374 + 198 = 572   265 + 247= 512

Why do I decompose 374, but not 198?

Answers may vary.

2. Add.

125 + 275 = 400

375 + 225 = 600

575 + 25 = 600

175 + 425 = 600

725 + 175 = 900

425 + 75 = 500

1. Add.

360	540	429	731
+ 280	+ 160	+ 150	+ 190
640	700	579	921
264	582	198	356
+ 139	+ 159	+ 278	+ 268
403	741	476	624

2. Add. Draw the arrows to show the best strategy.

44 + 23 + 18 + 27 + 16 + 42 = 170

55 + 12 + 21 + 36 + 15 + 19 + 18 + 24 = 200

3. Add or subtract. Hint: use Round-Up Strategy.

424 + 199 = 623

314 + 295 = 609

857 – 397 = 460

564 – 298 = 266

684 – 195 = 489

578 + 391 = 969

1. Unscramble each word to find out more about the ocean. Write the word in the squares.

istelho | h | o | s | t | i | l | e |

sidanls | i | s | l | a | n | d | s |

wsahlol | s | h | a | l | l | o | w |

acyl | c | l | a | y |

aes clmpte | s | e | a | | t | e | m | p | l | e |

1. Evaluate each expression.

88 – 39 + 22 – 44 + 58 = 85

39 + 32 – 46 + 27 – 38 = 14

58 + 19 – 49 + 15 – 27 = 16

63 – 37 + 45 – 12 + 24 = 83

2. Subtract. Hint: 524 – 70 – 5 = 524 – (70 + 5).

524	437	814	571
70	60	90	80
- 5	- 8	- 7	- 6
449	369	717	485
657	384	425	362
38	59	76	93
- 5	- 8	- 4	- 6
614	317	345	263

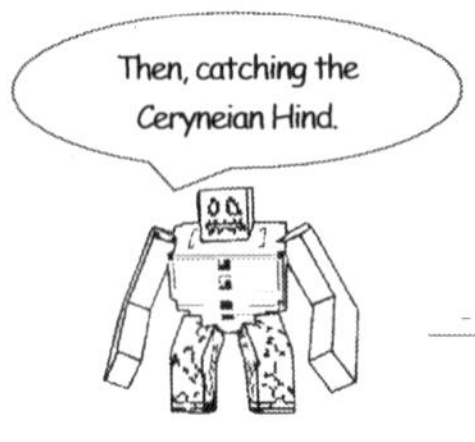

3. Write the missing numbers to make the sum and difference true.

	6	0	0		4	6	1
-	3	6	3	+	1	6	9
	2	3	7		6	3	0

1. Subtract.

520	640	800	700
- 225	- 375	- 365	- 425
295	265	435	275
813	485	680	612
- 276	- 268	- 483	- 209
537	217	197	403

2. Add. Hint: r 75 = 25 + 50.

The smartest Brainer thinks the easiest way is to calculate by 50's or 100's, so, make 50 or 100 with 25 and 75.

525 + 275 = 800 875 + 125 = 1000 25 + 775 = 800

425 + 375 = 800 225 + 175 = 400 625 + 175 = 800

175 + 325 = 500 425 + 575 = 1000 525 + 175 = 700

3. Subtract. Hint: Round-Up Strategy.

300 – 146 = 154

600 – 289 = 311

900 – 697 = 203

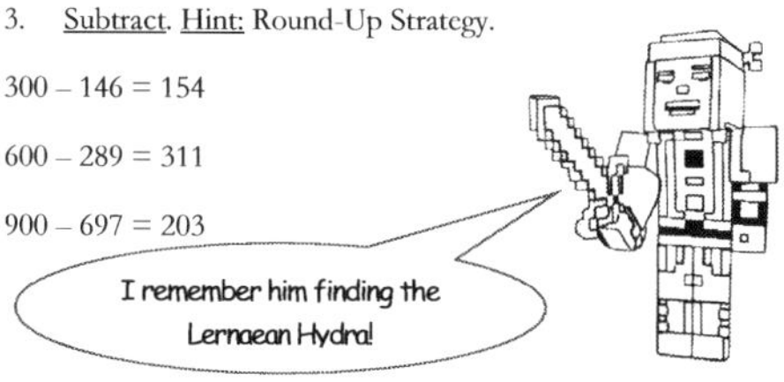

1. Add. Hint: decompose.

535 + 228 = 763 805 + 86 = 891 346 + 357 = 703

486 + 314 = 800 264 + 226 = 490 155 + 125 = 280

265 + 217 = 482 436 + 329 = 765 697 + 203 = 900

2. Subtract.

6 1	9 2	9 4	7 3	3 5
2 2	3 3	3 5	4 4	1 6
- 3 5	- 4 4	- 1 6	- 1 3	- 1 2
4	1 5	4 3	1 6	7
5 5	8 7	5 4	8 1	4 6
1 7	4 9	2 6	5 3	1 8
- 2 3	- 2 2	- 2 5	- 2 4	- 2 1
1 5	1 6	3	4	7
7 7	5 8	6 6	8 2	9 9
7	8	6	2	9
- 1 7	- 2 6	- 4 2	- 4 8	- 7 1
5 3	2 4	1 8	3 2	1 9

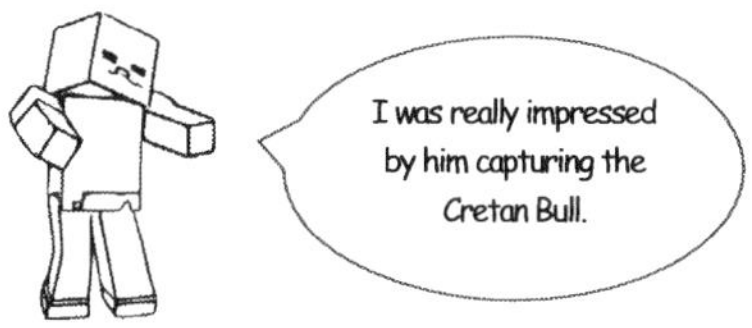

1. Subtract.

500	800	500	600
- 225	- 635	- 365	- 385
275	165	135	215
300	400	1000	1000
- 175	- 385	- 565	- 395
125	15	435	605

2. Add. Use 2 or 3 different addition Strategies.

386+155 = 155+400–14=386+14+141=380+120+6+35=541

548 + 244 = 792

653 + 197 = 850

365 + 477 = 842

1. Fill in the missing numbers to make the equations true.

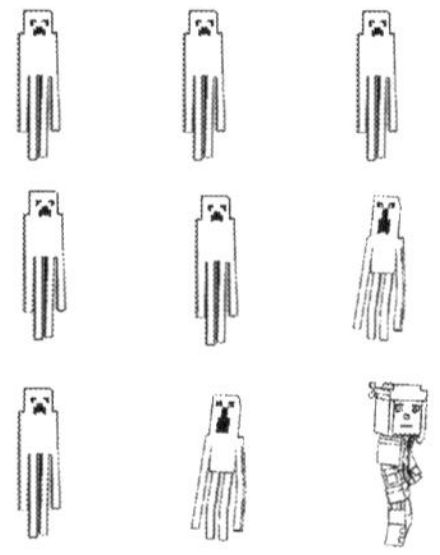

22 + 22 + 22 = 66

22 + 22 + 28 = 72

22 + 28 + 35 = 85

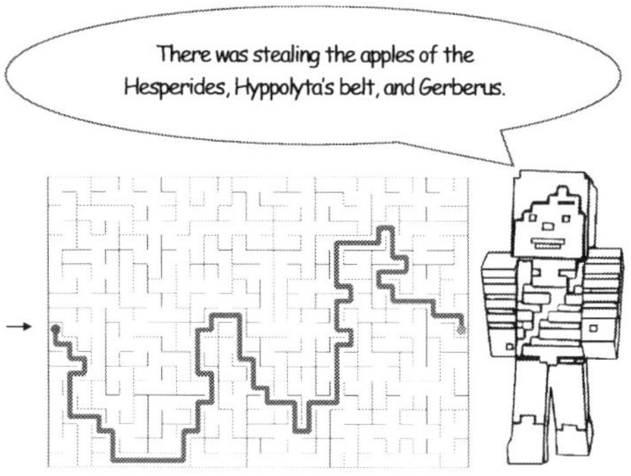

1. Write the missing numbers 275, 300, 650, 850, and 925 to make the equation true.

850 - 275 = 925 - 650 + 300

2. Answer the question.

A	B	C	D	E	F	G	H	I	J	K	L	M
1	2	3	4	5	6	7	8	9	10	11	12	13
N	O	P	Q	R	S	T	U	V	W	X	Y	Z
14	15	16	17	18	19	20	21	22	23	24	25	26

W h a t   d o   y o u   n e e d   t o
23 8 1 20   4 15   25 15 21   14 5 5 4   20 15

c r a f t   s i x t e e n   r a i l s?
3 18 1 6 20   19 9 24 20 5 5 14   18 1 9 12 19

One stick and six iron ingots.

1. Answer the questions.

If you build bricks in pairs, how many pairs will you make? 10

How many will be left? 0

How many 3s can you build with the bricks? 6

How many will be left? 2

How many 4s can you build with the bricks? 5

How many will be left? 0

How many 5s can you build with the bricks? 4

How many will be left? 0

How many 6s can you build with the bricks? 3

How many will be left? 2

How many 7s can you build with the bricks? 2

How many will be left? 6

How many 8s can you build with the bricks? 2

How many will be left? 4

How many 9s can you build with the bricks? 2

How many will be left? 2

How many 10s can you build with the bricks? 2

How many will be left? 0

Made in the USA
San Bernardino, CA
02 December 2019

60761143R00058